Field Manual
No 24-12

*FM 24-12
HEADQUARTERS
DEPARTMENT OF THE ARMY
Washington, DC, 17 July 1990

FM 24-12

COMMUNICATIONS IN A "COME-AS-YOU-ARE" WAR

This Page Intentionally Blank

Table of Contents

Preface

Purpose and Scope

This publication is designed as a handbook of helpful information and methods for providing essential communications support in the face of communications equipment shortages in AC and RC units. It also addresses methods for achieving interoperability between generations of tactical communications equipment.

This publication can be used by the planner, the supervisor, and the operator. It suggests field expedient solutions, most of which are found in current doctrinal literature. Some of these suggestions are field expedient techniques – methods not normally considered that may require additional items.

The employment techniques are described in general terms. Specifics will be dictated by the situation and by the appropriate equipment manuals. These techniques are proven and may even provide increased flexibility in communications systems. The recommended methods are not intended to be contrary to regulations or policies on maintaining, using, modifying, or cross leveling equipment. They do, however, provide innovation and flexibility in getting the message through on the modern battlefield.

A chapter addressing TRI-TAC equipment, especially DGM and the basis of how it interfaces with MSE will be added to this manual in the next revision. The appendix addresses initial lessons learned about MSE to TRI-TAC interoperability.

User Information

The proponent of this publication is HQ TRADOC. Your comments on this publication are encouraged. Submit changes for improving this publication on DA Form 2028 (Recommended Changes to Publications and Blank Forms) and key them to pages and lines of text to which they apply. If DA Form 2028 is not available, a letter is acceptable. Provide reasons for your comments to ensure complete understanding and proper evaluation. Forward your comments to Commander, United States Army Signal Center and Fort Gordon, ATTN: ATZH-DTL, Fort Gordon, Georgia 30905-5075

This Page Intentionally Blank

Chapter 1

Planning Considerations

1-1. General

a. The RC makes up nearly half of the military capability of today's US Army. RC units, in many instances, do not have a complete fill of authorized communications equipment; what they have may consist of different generations of equipment. Current economic realities limit how much additional communications equipment RC units can expect to receive. Thus, RC units must be prepared to mobilize with the equipment on hand in a "come-as-you-are" war.

b. AC units are also affected by this dilemma. Many RC units are related to an AC unit under an affiliation or roundout program. More specifically, they train and operate with their active counterparts. If old and new equipment must be interfaced, both AC and RC units must know the proper equipment interface procedures.

c. A similar problem could exist between allied units using different types of communications equipment, and US units using standardized, modernized, or upgraded communications equipment. Any military unit faces the possibility of operating with a shortage of equipment. Combat losses, excessive usage, maintenance problems, normal wear and tear, and delayed receipt of new equipment reduce equipment availability.

d. OPSEC principles must be inherent in all phases of a "come-as-you-are" war. The principal OPSEC elements of physical security, information security, signal security, and military deception must be continually applied, not only during combat operations, but also during peacetime planning. This ensures the protection of military operations and activities and prevents hostile exploitation of identified weaknesses. Shortages of equipment and personnel, equipment interface problems, training deficiencies, and other such problems are exploitable weaknesses that must be properly protected. Remember, the way we practice is the way we fight.

e. The NBC environment must be included in planning considerations. Communications in NBC conditions must be realistically anticipated and discussed with candor.

1-2. Reduced Equipment Planning

a. The key to operating with reduced equipment quantities is advance planning and action. Viable alternatives must be devised and supporting equipment and personnel must be requisitioned, obtained, and readied. Critical questions must be answered. For example, what communications support can the first brigade expect based on the division's present capabilities? What is the tactical situation and what actions are planned next? What signal assets are available for supporting these actions? What are the minimum necessary communications support requirements for DTAC to the brigade TOC, DTOC, DISCOM, or others? What are the priorities?

b. There are no pat answers. We do know that the following types of traffic are essential:

- Command.
- Operations.
- Intelligence.
- Fire support.
- Logistics.

c. The next question is which means will be allocated to the respective critical needs? FM radio is used for the immediate command, operations, and fire support traffic. Some of this traffic will have to be passed over AM radio and multichannel radio systems, supplemented by alternate means. Lesser priority traffic should be passed over alternate means such as motor and air messenger service to the maximum extent possible.

1-3. NBC Environment

a. In the past, combat communications have been installed under difficult yet understandable conditions such as bad weather, limited equipment, and even hostile fire. These conditions are understandable because they have been experienced. We have grown up in good and bad weather. We have read combat histories, watched combat films, and even listened to soldiers who have participated in combat. The future battlefield will include an NBC environment not yet experienced. (See FM 3-100 for the fundamentals of NBC defense.)

b. The equipment may be contaminated by biological and chemical agents. Decontamination of internal electronic components may be difficult if not impossible. (See FM 3-5 for NBC decontamination and FM 3-3 for NBC contamination avoidance.) Thus, continuous operations in MOPP and its effect on personnel and installations must be included in planning estimates. Forward communications teams, PCM relays, and FM retransmission stations need to become familiar with displacement under limited visibility while in MOPP.

c. Operators are not as effective while in MOPP. Handling knobs while wearing bulky gloves can frustrate operators. Voice communications are difficult, not only with FM and AM/SSB radios, but also with orderwires, switchboards, and patch facilities. The problems in understanding verbal instructions can slow system installation and subscriber use. The MOPP equipment can generate heat and cause operator sweating which irritates the soldier causing anxiety and inattention to details. (See FM 3-4 for individual and collective NBC protection.)

d. Leaders, especially NCOs and first line supervisors, must understand that direct involvement in this situation may not solve the problem. The universality of MOPP appearance and the distortion of voice quality make familiar leaders appear unrecognizable. Only through proper supervisor training can soldiers' natural apprehension be translated into confidence.

1-4. NBC Communications

a. Communications will be affected in at least two ways during nuclear warfare: communications blackout and physical damage to equipment.

    (1) Communications blackout is caused by intense ionization of the atmosphere in the vicinity of the blast. The blackout may last for a few seconds or several hours. It may be more severe on some frequencies than on others. During this period, communications is impossible.

    (2) Physical damage depends on the nearness of the blast to the equipment. It can range from total disintegration (at ground zero) to thermal (heat) damage (several miles away) to electrical breakdown caused by the EMP radiated from the burst (several miles away). The EMP is perhaps the most subtle cause of physical damage. It is a very intense radio wave of extremely short duration produced at the instant a nuclear weapon is detonated. It usually lasts a fraction of a second. But the power it may deliver to a radio receiver can be a billion times greater than what is normally received from a transmitter. This extremely high power density can damage some signal equipment. EMP is silent and invisible.

b. The ability of a unit to continue to communicate during a tactical nuclear war will depend on planning, training, and equipment hardness. These actions must begin long before the war begins. System planning must use minimal resources to perform the mission allowing a portion to standby. Training must incorporate the use of hardened CPs and EMP prevention steps, such as shielding by natural terrain, burying of cables, and disconnecting equipment when not in use. Equipment hardening and buffering devices are included in new equipment development. However, these steps are based on mission accomplishment in a nuclear environment. All soldiers must train the way we expect to fight and communicate.

Chapter 2

FM Radio Operations

2-1. General

a. Single-channel FM voice radios are the primary communications means used in almost all tactical Army units below brigade level. FM radios give the tactical commander quick, reliable, and flexible communications needed to control the battle. The AN/VRC-12 series used by active and reserve units is the only family of FM radios currently used by the Army. With the fielding of SINCGARS, the Army may use two families of FM radios. The signal personnel of both AC and RC must prepare to operate tactical FM radio nets containing both families of radios.

b. FM radios must take some of the additional burden when shortages of multichannel equipment force reliance on other means of communications. FM radios add a great deal of flexibility to our communications system. This section addresses techniques useful in providing essential command and control communications in the face of equipment shortages.

2-2. Frequency Planning

a. Figure 2-1 shows a comparison of frequency ranges between the AN/VRC-12 series and the SINCGARS radio sets. In addition to the extended range of SINCGARS, channel-spacing problems must be anticipated when interfacing AN/VRC-12 and SINCGARS. The channel spacing for AN/VRC-12 is 50 kHz. The channel spacing for SINCGARS is 25 kHz. When interfacing, frequencies must end in 00 or 50.

SINCGARS	AN/VRC-12
30.000	30.000
30.025	
30.050	30.050
30.075	
30.100	30.100
30.125	
30.150	30.150
30.175	
30.200	30.200

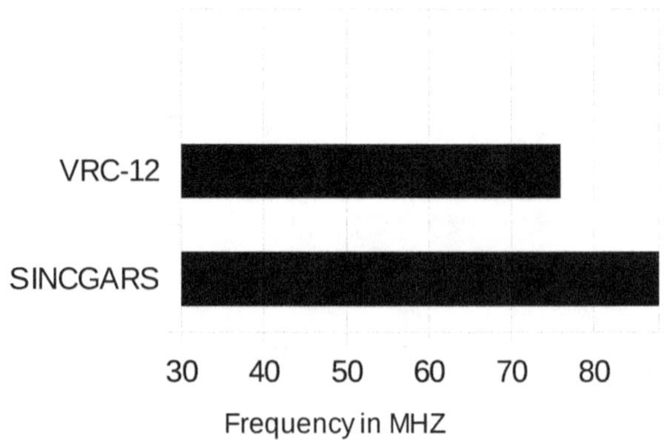

Figure 2-1. Frequency comparison chart

b. Obviously, all radios for a particular net must be capable of operating on the same frequency. Net frequencies must be assigned with primary consideration given to the old radio's frequency tuning capabilities. This applies also to channel spacing. The AN/VRC-12 series has a channel every 50 kHz and the SINCGARS equipment every 25 kHz. Nets involving both series radios must consider these differences and plan for proper channel separation.

2-3. Planning Range

a. Older radios have less range than the newer radios, so the maximum planning range must not exceed that of the older radios. Antennas on vehicles are vertically polarized; therefore, polarization usually presents no problem. The distance problem may be eased by the careful placement of retransmission stations in the unit's area of operations. Retransmission is effective but must be carefully controlled and properly employed using electronic warfare considerations.

b. The range of the FM radio sets can be extended by the proper location and orientation of the antenna system in regard to the vehicle and terrain. Additional distance can be obtained using elevated ground plane antennas such as the RC-292 or OE-254. Field expedient antennas may also be used not only to increase range but also to provide more directivity while reducing interference and detection. The SINCGARS may use the RC-292 in the single-channel mode. The OE-254 may be used in single-channel and frequency-hopping modes. Field expedient antennas may be used with SINCGARS in single-channel mode.

c. The planning range can be further extended with retransmission operations.

> (1) Retransmission, or retrans for short, offers the commander a valuable alternative when multichannel equipment is in short supply or absent. As with NRI, retransmission is often not used to maximum advantage because of lack of knowledge or lack of confidence in its effectiveness. A shortage of multichannel equipment requires better planning and use of all other communications assets; retransmission is no exception.

> (2) A primary application of retransmission is the extension of a particular communications link such as an FM command net or a fire direction net. Another application might be a logistical link from brigade trains to the DISCOM area in the absence of multichannel. Traffic on this link would be for urgent requests for resupply of critical items only such as ammunition or POL, or for a contact team for the maintenance of critical items. Routine traffic should be sent by other means.

> (3) Retransmission should also be used to support anticipated operations or planned moves of important elements. For example, if a brigade CP is moving to a certain location at 1600 hours, a retransmission station could provide a communications link back to the DTOC. Retransmission may also allow the brigade CP to locate in a position that provides better physical security while still maintaining its essential FM radio communications.

> (4) Retransmission sites must be carefully chosen to maximize retransmission distance while at the same time minimizing enemy interception. Several alternate sites should be chosen for each retransmission facility to allow for periodic displacement.

d. HF radios, such as the AN/GRC-106 or the AN/PRC-104, are used for long-range communications.

2-4. FM Radio Security

a. The AN/PRC-25 is not capable of operations using secure equipment. The AN/PRC-77 and AN/VRC-12 series radios can be secured with VINSON. SINCGARS can also be secured with VINSON. Planners should attempt to exchange equipment if certain nets must be on-line secured, and leave other nets to use low-level encryption and authentication procedures. Regardless of which security system is used, all nets must use proper radio procedures.

b. Chapter 7 discusses COMSEC. It covers the use of proper procedures and encryption systems in voice and record communications. Part of our security effort must be directed at operations that prevent the enemy from locating our emitters or analyzing our traffic. This is essential to survival on the modern battlefield. Every operator and user of signal systems should read and practice the techniques described in FM 24-33 and ACP 125(D). All communicators should practice daily the basic ECCM techniques described below.

(1) Use the lowest power possible for the required communications when power settings are adjustable. This is especially important the closer the transmitter is to the FLOT.

(2) Reduce on-the-air communications time. Both the quantity and length of transmissions must be kept to a minimum to deny the enemy the opportunity to detect and exploit friendly communications. Minimal transmissions should be coupled with frequent moves for greater security against enemy direction finding efforts.

(3) Change call signs and frequencies and use the proper authentication and COMSEC practices as specified in the unit SOI.

(4) Train all radio operators to practice sound radiotelephone procedures and use them in all training and operations.

(5) Emplace antennas and all noncommunications emitters properly. Terrain masking is an invaluable technique for denying the enemy knowledge of your location and unit.

(6) Use digital communications terminals such as the AN/PSC-2 or TACFIRE digital message device when possible to take advantage of burst and error corrected transmission.

c. Many RC units are equipped with the AN/VRC-47 radio, enabling them (doctrinally) to operate in two separate radio nets at the same time. When the AN/VRC-47 is equipped with a KY-38 FM security device (or similar secure equipment), the RT-524 receiver-transmitter continues to function. But the R-442 auxiliary receiver will be inoperative since the KY-38 accommodates only the RT-524. The result is one AN/VRC-47 that was formerly engaged in a two-net function is now capable of only single-net operations. This shortfall must be considered in the planning. Similar problems occur when installing the KY-38 on the AN/VRC-44 and AN/VRC-48. If using the VINSON KY-57 security device, both the R-442 and RT-524 will operate in the secure mode.

2-5. Squelch Capabilities

a. The AN/VRC-12 series radios have the ON and OFF positions in both new squelch and old squelch. The old squelch was used with the AN/GRC-3 through -8 series radios which are no longer in the inventory. The AN/PRC-77, AN/PRC-25, and SINCGARS can be operated in new squelch or without squelch. The AN/VRC-12 series radios should be used in new squelch on or new squelch off mode only.

b. Squelch is particularly important when netting (interoperability) with allied forces whose tactical FM radios operate in the old squelch and have limited frequency range. Close coordination is required when netting with allied forces.

2-6. Net Radio Interface

a. NRI is a highly effective method for bridging the commander's two primary means of command and control: tactical radio and telephone networks. It is normally used only by commanders and key staff members, but in a "come-as-you-are" war more people may need to use NRI. This is because shortages of multi-channel equipment force the commander to find alternate communications routes, and NRI is one of the most flexible.

b. Commanders have not made full use of their NRI capabilities in the past because-

- Some commanders and communications personnel do not know enough about NRI.
- Some units lack technical expertise, resulting in lack of confidence by the commander in the NRI system.
- Some commanders do not trust vital communications traffic to NRI systems because of lack of confidence in NRI.
- NRI, at this point in time, is not capable of end-to-end encryption.

c. NRI extends communications distance because it connects the tactical radio into the division/corps wire system. Some NRI stations can also be used as retransmission stations, not simultaneously, but alternately in either mode. For example, the commander in his vehicle calls the NRI station to place a call back to the DTOC. The NRI operator attempts to call through, but the circuits are busy. As a result of thorough training and by knowing the SOI retransmission frequency, the operator switches over to the retransmission frequency and puts the call through to the NRI station serving the DTOC.

d. NRI stations, like other FM radio installations, must be moved periodically to various alternate sites, both to adequately serve the headquarters or elements it supports and to enhance security and survival. The move may be necessary to support a new tactical CP or to support fast-moving operations. These moves are supported by--

- Planning acceptable communications sites.
- Considering the mission and security requirements.
- Planning and installing wire/cable to tie in to the division or corps wire/cable system.
- Establishing a "jump" facility that moves into position to support operators before shutting down and moving to another facility.

Pre-positioning of wire/cable system terminations is absolutely necessary on the fast-moving battlefield.

e. NRI systems must connect to switchboards to have access to the telephone network. The switchboards can be manual or automatic boards and either can process NRI calls--but not without prior identification of the NRI circuits and adequate training of operators (both switchboard and radio). Switchboard operators must know telephone traffic diagrams well to react adequately and quickly. SOPs must identify those individuals who are authorized to use NRI circuits. All users of NRI systems must use low-level security procedures since NRI is not secure. NRI frequencies are found in the unit SOI. NRI procedures for both radio and telephone are found in the supplemental instructions in the unit SOI. All NRI users must use proper procedures when communicating.

f. The AN/GSA-7, AN/GRA-39, AN/GRA-6, and C-6709/G are associated items of NRI operation. More complete explanations are covered in FM 24-18.

(1) Radio set control AN/GSA-7 provides an interface between a radio set and a switchboard which can be located for planning purposes up to 16 km (10 miles) from the radio set. There are four methods of providing NRI (Figure 2-2), depending on the number of AN/GSA-7s in the system. The four methods are described in FM 24-18. These variations provide for both attended and unattended operation of the AN/GSA-7.

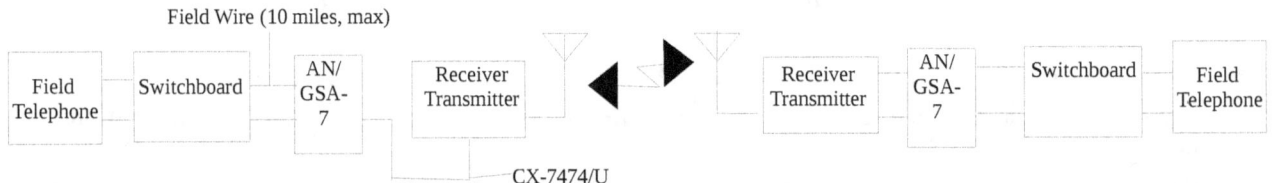

1. AN/GSA-7 at Both Stations;Radio Operator not Required at AN/GSA-7.

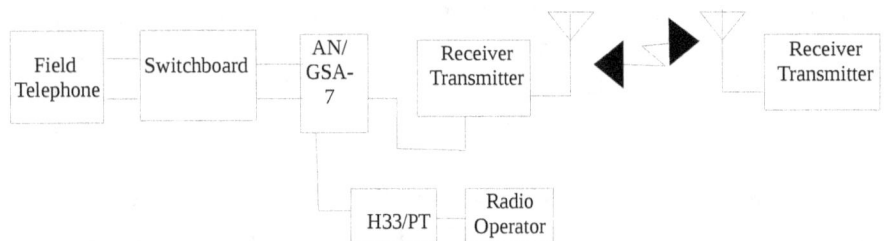

2. AN/GSA-7 at one station: Radio Operator required at AN/GSA-7.

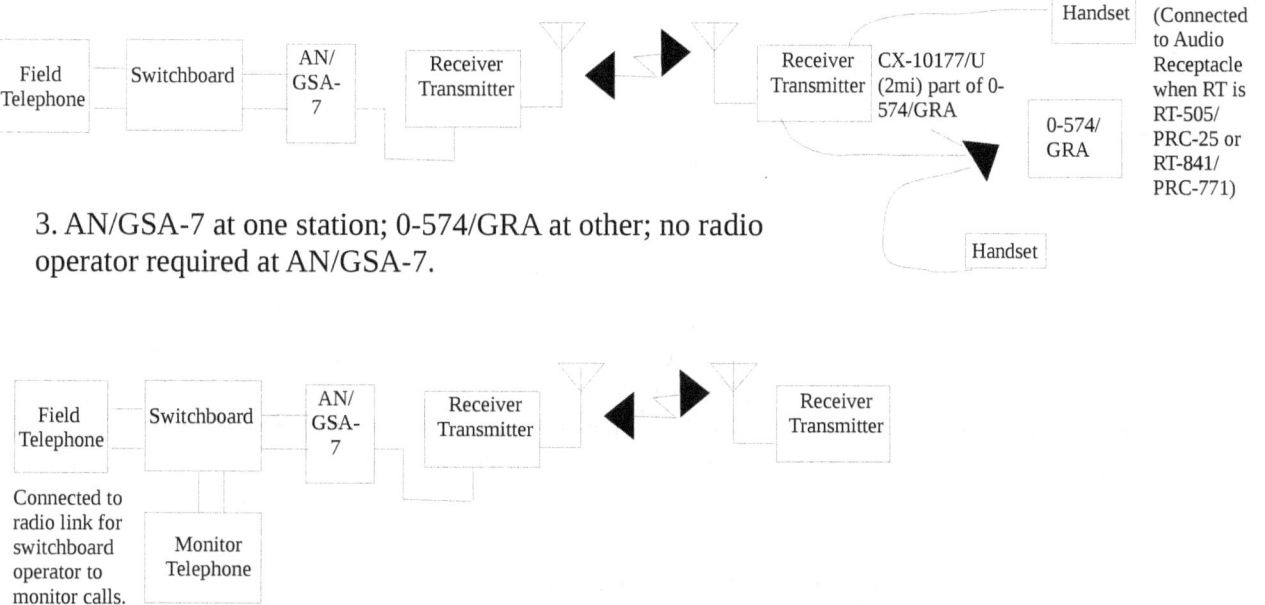

3. AN/GSA-7 at one station; 0-574/GRA at other; no radio operator required at AN/GSA-7.

4. AN/GSA-7 at One Station; Radio Monitor Telephone. No Radio Operator Required.

Figure 2-2. Four methods of NRI.

(2) Radio set control group AN/GRA-39 can provide remote control of receiver-transmitters up to approximately 3.2 km (2 miles). It can also provide NRI among SB-22, SB-86, SB-3082, and SB-3614 switchboards and receiver-transmitters such as the AN/PRC-25, AN/PRC-77, and AN/VRC-12 series radios. NRI operation with the AN/GRA-39 can be remoted up to 1.6 km (1 mile) between the switchboard and the receiver-transmitter. Specific procedures for remote operation and NRI operation are described in the operator's technical manuals for the radios and for the AN/GRA-39.

(3) Radio set control AN/GRA-6 can provide remote radio control of a radio set up to 3.2 km (2 miles). It can also be used to provide NRI between an SB-22 switchboard and the radio set.

(4) Radio set control C-6709/G provides the capability for manned integration between 4-wire tone signaling telephone communications systems and push-to-talk radio systems. The C-6709/G is compatible with both current and future wideband transmission requirements. The 300 Hz to 70 kHz baseband allows the unit to accommodate a wide variety of interfaces for data communications and other needs. Radio keying can be accomplished with manual control by the NRI operator, by DTMF procedures, and/or by automatic voice actuation. It contains an H-250 handset, an H-325 headset microphone, and connecting cables with a basic unit. The C-6709/G provides facilities for controlling transmitter/receiver circuits of a variety of tactical radios in a 4-wire switched system comprised of AN/TTC-38s, AN/TTC-39s, SB-3614(V), CNCEs, and radios with COMSEC, such as PARKHILL and VINSON.

2-7. FM Radio Operations Example

a. You are the S3 of the 52d Division (Mechanized) Signal Battalion. The DTAC is presently in the vicinity of the 1st Brigade. The G3 has notified you that the tactical CP is moving to the vicinity of the 3d Brigade and will be on site in four hours. No multichannel equipment is available to support the tactical CP or the 3d Brigade Headquarters. The commanding general's M577 has two FM radio sets, but only one RATT set is available for support at the new site.

b. How can you provide for minimum essential support for the tactical CP at the displacement location within four hours?

c. Obviously you cannot provide the tactical CP with the full doctrinal communications system at the displacement location. First, examine how the two FM sets could be employed. One set will be operating in the division command net. The other FM set must be used for an NRI circuit back to the extension node serving the 3d Brigade. In this situation, at least two priority circuits must be engineered from the extension node to the DTOC. These circuits would go to the G2/G3 and FSE in the DTOC. This configuration allows the CG to use his priority NRI to establish calls to the appropriate element in the DTOC. As a minimum-

- Extension node NRI stations supporting the 3d Brigade must be reassigned and must be prepared to support the CG's high priority calls.
- Dedicated circuits must be engineered from the extension node switchboard through the division communications system to the DTOC.
- NRI and switchboard operators must be well briefed for this type of system.
- DTOC personnel must be prepared to share the two priority lines from the extension node.

d. A retransmission station (if available) is needed to support the NRI system back to the extension node when actions to accomplish the above begin. Also, field wire lines must be installed (terrain and tactical situation permitting) to provide additional circuits to the new tactical CP location from the extension node. These wire circuits need a way that permits the NRI system to leave the air. This reduces the electronic signature of the tactical CP and improves its survivability. Also, the one RATT set at the tactical CP must be used for both command and operations traffic.

2-8. Technical Characteristics

FM operations are the backbone of combat communications. To plan communications networks effectively, the planner must know the technical characteristics of the radio equipment. Table 2-1 compares the important characteristics of the receiver-transmitter units that are the chief components of both the AN/VRC-16 series and the SINCGARS FM radios. Especially important are frequency, channel spacing, squelch, and secure equipment capability.

Table 2-1. Charactaristics of both old and new generation FM radio equipment.

RT Unit	Frequency Coverage (MHz)	Tuning	Channel Spacing	Power Output (Watts)
RT-524	30-75.95	Detent	.05 MHz	35-85 High 0.5-10 Low
RT-246	30-75.95	Detent	.05 MHz	35-85 High 0.5-10 Low
RT-505 (PRC-25)	30-75.95	Detent	50 kHz	1.1-2
RT-841 (PRC-77)	30-75.95	Detent	.05 MHz	1.5-4
SINCGARS	30-87.975	Synthesized	25 kHz	(low) 500 microwatts (Medium) 160

	Range (mi)(km)	Squelch	Using Unls	microwatts (high) 4 watts Security Equipment
RT-524	25mi (40km) High 5mi (8km) Low	New/Old/Off	All	KY-8/KY-38/KY-57
RT-246	25mi (40km) High 5 mi (8km) Low	New/Old/Off	All	KY-8/KY-38/KY-57
RT-505 (PRC-25)	5mi (8km)	New/Off	All	None
RT-841((PRC-77)	5mi (8km)	New/Off	All	KY-38/KY-57
SINCGARS	0.300m (low) 0.3-4km (medium) 4-16km (high) W/ power amperes (AN/VRS-89,-90,-91,-92) 6-35km	New/Off	All	KY-57

2-9. Typical Configurations

Current typical configurations of FM radio equipment are listed in Table 2-2.

Table 2-2. Configurations of AN/VRC-12 series FM radios.

AN/VRC-12 Family								
Configurations								
Component (VRC)	12	43	44	45	46	47	48	49
RT-246/VRC	1	1	1	2	0	0	0	0
R-442/VRC	1	0	2	0	0	1	2	0
RT-524/VRC	0	0	0	0	1	1	1	2
AT-912/VRC	1	1	1	2	1	1	1	2
O/AS-1729								

Chapter 3

AM Radio Operations

3-1. General

Equipment shortages or differences may cause serious problems. The problems could result in having only one or two radio sets to pass all the traffic normally passed by six or more sets. AM radios, combined with ancillary equipment, can pass traffic in either of three modes: voice, RATT, or CW. This chapter addresses solutions to the problems caused by equipment shortages or differences.

3-2. Reduced Assets

Following are alternatives for operating with reduced assets:

a. Operate one AM radio in several nets using an established time schedule. Select the most important net to monitor and operate in. Enter the other nets only when necessary to pass traffic. Enter the other nets at preplanned times or notify the other stations by telephone or FM radio at unscheduled times. The schedule should change every day and be randomly generated to preclude the enemy from analyzing your traffic pattern.

b. Preplan all messages by using brevity lists and codes to shorten the time spent on the air.

c. Use the TCC's off-line teletypewriters to prepare teletypewriter tapes prior to submitting traffic to the RATT operator for transmission. This reduces the RATT operator's burden and saves transmission time.

d. Use FDX operation on equipment which has FDX capability. Much more traffic can be passed over FDX circuits than over half-duplex circuits, thereby reducing time required for passing traffic.

e. Transmit low priority traffic over alternate means, such as messenger or multichannel radio, if they are available.

f. Use one radio, if possible, for several individuals, staff sections, or units.

g. Establish a wire link with a distant station using the existing teletypewriter and secure device along with a telegraph terminal TH-5/TG or TH-22/TG when the radio or modem of a RATT system is defective. Speech-plus can also be provided using this technique by using telegraph-telephone terminal AN/TCC-14 or AN/TCC-29.

h. Use error correcting burst communication devices, such as the AN/PSC-2, if available, to cut down on air time and errors.

3-3. Equipment Considerations

a. When different types of AM radios must work together in the same net, the SOI must include proper frequency assignments compatible to each type of radio equipment. All nets using two or more different AM radios are restricted to certain frequency ranges or modes of operation. (The various technical characteristics of all the AM radios currently in the Army inventory will be covered later in this chapter.) Frequency and mode assignments must be coordinated prior to joint operations when units with different AM radios may be involved together.

b. When planning nets, radio planning ranges must be considered. Certain AM radios have much more power than others, so planning ranges must be based on the least powerful radio's capabilities. Related to the distance factor is the type and polarization of antennas. Antennas must be properly polarized and correctly oriented. For extended ranges, a half-wave doublet antenna, such as the AN/GRA-50, should be used whenever time and terrain permit.

c. By obtaining prior approval, various civilian radios with AM, CW, and SSB capabilities can be used. When using civilian radio equipment, proper military procedures will be used. For operation of ranges between 0 and 450 km, the near vertical incidence skyware technique described in FM 24-18, Appendix N, should be used. This will allow skip zone free omnidirectional communications at low power under all conditions.

> NOTE: Under no circumstances will citizen's band procedures be used.

3-4. AM Radio Security

Some military AM radios may be secured in the voice mode using the KY-65. Most AM radios can be secured for RATT operation using TSEC/KW-7 or TSEC/KG-84A security devices. Some of the older radios have not been modified to accept the KW-7 at this time but can be altered to accept it as required. If time and situation permit, ensure that all your radios have been modified to accept security devices. When operating a nonsecured radio in voice or CW modes, codes or off-line encryption methods must be used.

> NOTE: The KG-84A will only operate with AN/GRC-142/122 RATT sets equipped with MK-2488 installation kits.

3-5. Netting Old and New Equipment

Most AC use the newer families of SSB radios, whereas RC have a combination of the older AM and the newer SSB equipment. The new equipment can work with the older equipment, but it takes just a little extra care to make it work correctly. The planner needs to know the technical characteristics of all the radio sets involved to plan the communications network properly. Table 3-1 is a comparison of technical characteristics of the AM radios in the Army inventory.

Table 3-1. AM radio technical characteristics.

	Nomenclature	Configurations	Types of Service	Tuning	Power Output	Planning Range
OLD	AN/GRC-19	GRC-46, VRC-29, VSC-1	DSB Voice, CW, RATT	Continuous	100W	80km (50mi)
	AN/GRC-41		DSB Voice, CW	Continuous	Voice 400W CW/RATT 450W	1500km (1000mi)
	AN/GRC-26D		DSB Voice, CW, RATT	Continuous	Voice 400W CW/RATT 450W	1500km (1000mi)
	AN/GRC-109		DSB Voice, CW	Continuous	10W 15W	121km (75mi)
	AN/GRC-87	VRC-34	DSB Voice, CW	Continuous	Voice 7W CW 15W	16 to 50km 21 to 31 mi
	AN/GRC-93		SSB Voice, CW	Continuous	100W 1000W (w/AM-39/B)	80km (50mi)
	AN/MRR-8	(Receive Only)	DSB Voice, CW, RATT	Continuous	NA	160km (100mi)
	AN/MRT-9			Continuous	Voice 400W CW/RATT 450W	160km (100mi)
New	AN/GRC-106	GRC-122, GRC-142, VSC-2, VSC-3	AN/SSB Voice, CW, RATT FSK/NSK	Detent	Voice 400W CW-RATT 200W	80km (50mi), 160 to 2400km (100 to 1491mi)
	AN/PRC-47		SSB Voice, CW	Detent	High 100W Low 20W	
	AN/PRC-46A		DSB Voice, CW	Detent	Voice 1.5W CW 5W	24km (15mi)
	AN/PRC-70		SSB Voice, CW	Detent	Voice 7.5W CW 30W	121km (75mi) 4000km (2500mi)
	AN/PRC-74	PRC-74A, PRC-74B, PRC-74C	SSB Voice, CW	Detent	15 W	40km (25mi)
	IHFR AN/PRC104B	AN/GRC-213A, AN/GRC-193B	USB/DSB Voice, Data/RATT FSK, *CW	Detent	AN/PRC-104 20W AN/GRC-213 20W AN/GRC-193 100-400W	2385km (1491mi)

*Note – IHFR only operates CW in non-STAT mode.
Note – AN/PRC-104 is User Owned and Operated Equipment.

a. The most important technical characteristics to consider when netting two different radios are the type of tuning and type of emission.

(1) Table 3-1 shows that the older radios have continuous tuning whereas the newer radios have detent tuning. The difference between these two is that detent-tuned radios can tune only to certain frequencies and cannot tune to the in-between frequencies to which the continuous-tuned radios can tune. The continuous-tuned radios must tune to the detent-tuned radios. This includes the radio systems used by other military services (such as Marines, Navy, Air Force) . Continuous-tuned radios operate approximately 1.5 kHz below the detent-tuned radio's frequency. The NCS should have a new series radio set to which all radios can tune. If the NCS does not have a new series radio set, the operator should direct a station with a new series radio

to provide the signal to which all others tune. Check Table 3-2 for compatible frequency ranges.

Table 3-2. AM frequency ranges.

Nomenclature	Frequency Range	Frequency in MHz							
		0	5	10	15	20	25	30	35
AN/GRC-19 Transmit	1.5 to 20.0 MHz	xx	xx	xx	xx	xx			
AN/GRC-19 Receive	0.5 to 32.0 MHz	xx	xx	xx	xx	xx	xx	xx	x
Transmitter T-368 *	1.5 to 20.0 MHz	xx	xx	xx	xx	xx			
Receiver R-390 **	0.5 to 32.0 MHz	xx	xx	xx	xx	xx	xx	xx	x
AN/GRC-109 Transmit	3.0 to 22.0 MHz	xx	xx	xx	xx	xx	x		
AN/GRC-109 Receive	3.0 to 24.0 MHz	xx	xx	xx	xx	xx	x		
AN/GRC-87	2.0 to 12.0 MHz	xx	xx	xx	x				
AN/FRC-93	3.4 to 5.0 MHz 6.5 to 30.0 MHz	Xx xx	xx xx	 xx	 xx	 xx	 xx	 xx	
AN/GRC-106 AN/GRC-106A	2.0 to 29.999MHz 2.0 to 29.999MHz	Xx xx	Xx xx	Xx xx	Xx xx	Xx xx	xx xx	Xx xx	
AN/PRC-47	2.0 to 11.999MHz	xx	xx	xx	x				
AN/PRC-64A	2.2 to 6.0MHz	xx	xx	x					
AN/PRC-70 ***	2.0 to 75.999MHz	xx	xx	xx	xx	xx	xx	xx	xx
AN/PRC-74/74A	2.0 to 11.000MHz	xx	xx	xx	x				
AN/PRC-74B/74C	2.0 to 17.999MHz	xx	xx	xx	xx	x			
IHFR AN/PRC-104, AN/GRC-213, and AN/GRC-193	2.0 to 29.9999MHz	xx	xx	xx	xx	xx	xx		

* T-368 is a component of AN/GRC-26D, AN/GRC-41, and AN/MRT-9
**R-390 in a component of AN/GRC-26D, AN/GRC-41, AN/MRR-8, and AN/MRT-9
***AN/PRC-70 is FM from 30.00 to 75.999MHz

(2) Emission types must match. In the CW mode, the type of emission for both old and new series radio sets is the same. In the FSK mode, the type of emission is the same, but the way the carrier shifts is different -a problem that can be overcome. The type of emission for voice, however, is different. The old series equipment uses DSB while the new series equipment uses SSB and compatible AM. Only the compatible AM mode of the new series radio can be used with the older equipment. SSB cannot be used to communicate with the older series equipment.

b. The difference in frequency ranges must also be considered when operating old and new series radio sets together. The old series radio sets have a transmitting frequency range between 1.5 to 20 MHz, and a receiving frequency range between 0.5 to 32 MHz. The new series radio sets have a frequency range for transmitting and receiving from 2.0 to 29.999 MHz. When operating between the old and the new series radios, the operating frequency must be within 2.0 to 20 MHz.

c. The most commonly used old series RATT sets are the AN/GRC-46 and the AN/GRC-26D. They are used by RC. The most commonly used new series RATT sets are the AN/GRC-142 and the AN/GRC-122. They are used by both RC and active Army. Older generation radio sets may be used by other services, and may be encountered during joint operations. Characteristics of both old and new series equipment are listed in Table 3-3.

Table 3-3. Characteristics of old and new generation AM radio nets.

	Old Family	New Family
RT Unit	AN/GRC-19 T-195 R-392 AN/GRC-26 T-368 R-390 (2 Each)	AN/GRC-106 RT-662 or RT-834 (AN/GRC-106A) AM-3349 IHFR AN/PRC-104/GRC-213 AN/GRC-193
Frequency Coverage	Transmit 1.5-20.0 MHz Receive 0.5-32.0 MHz	Transmit and Receive 2-29.999MHz (RT-662/GRC) 2-29.9999MHz (RT-834/GRC)
Tuning	Continuous	Detent
Power Output	T-195 100W T-368 450W	200 watts RATT/CW 400 watts (pep) Voice
Channels	NA (Continuous Tuning)	Every 1 kHz (RT-662) Every 100Hz (RT-834)
Modes of Operation	DSB Voice CW FSK	SSB Voice, Compatible AM Voice NSK (85Hz shift) FSK (850Hz shift) CW
Security Equipment	*KW-7 (RATT only)	*KW-7 (RATT only)
*TSEC/KG-84A to replace TSEC/KW-7.		

3-6. Interoperability Procedures

The rules for netting old and new series radio sets for voice, CW, and RATT operations are given below.

a. Voice operations.

    (1) For new series radio sets An/GRC-142/122

- Tune for normal voice operation.
- Change the SERVICE SELECTOR switch on the RT-662 or RT-834 from the AM position.
- Conduct normal voice operation.
-

    (2) For old series radio sets AN/GRC-46 and AN/GRC-26D--

- Tune equipment as usual for voice operation.
- Rotate the CONTINUOUS TUNING dial for clearest voice reception while receiving voice signal from a station in the net using a new series radio set.
- Realign transmitter to receiver.
- Conduct normal voice operation.

(3) For voice tuning procedures of old and new radios, see Table 3-4.

Table 3-4. Voice tuning procedures.

Old Radios	New Radios
Tune equipment as usual.	1. Tune for SSB voice operations. 2. Change the SERVICE SELECTOR switch from SSB/NSK to the compatible AM position

b. CW operations.

(1) For new series radio sets AN/GRC-142/122--

- Tune radio set for normal operation.
- Change the SERVICE SELECTOR switch on the RT-662 or RT-834 to the CW position.

NOTE: In the CW mode, the transmitted RF signal is 2 kHz higher than the frequency indicated on the receiver-transmitter MC and KC controls.

- Lower the operating frequency by 2 kHz on the RT-834/662. Key radio set with CW keying device and adjust BFO control left or right for comfortable listening tone.
- Conduct normal CW operation..

(2) For old series radio sets AN/GRC-46 and AN/GRC-26D--

- Tune radio sets for normal CW operations.
- Rotate the CONTINUOUS TUNING dial on the receiver until a clear CW signal is heard while receiving a CW signal from a station in the net using a new series radio set.
- Realign transmitter to receiver.
- Conduct normal CW operations.

(3) For CW tuning procedures of old and new radios, see Table 3-5.

Table 3-5. CW tuning procedures.

Old Radios	New Radios
1. Tune equipment as usual. 2. Lower the operating frequency by 1.5kHz if using AN/FRC-93	1. Tune equipment for CW operation. 2. Lower the operating frequency by 2kHz.

c. RATT operations.

NOTE: Other than netting with the continuous tuning dial for voice and CW operations, the operator of the new series radio set was the only one that had to make changes in the normal tuning procedures of his radio set. For RATT operations, both the operator of the old series and the new series radio sets must make changes from the normal tuning procedures. The primary reason is the position of the mark and space signals in relation to the carriers of the two types of equipment. Old series radio sets transmit the mark signal above the carrier and the space signal below the carrier. The new series radio sets transmit the mark signal below the carrier and the space signal above the carrier.

(1) For new series RATT sets AN/GRC-142/122, VSC-2, and VSC-3--

- Tune RATT equipment as usual for normal RATT mode of operation (Table 3-6).
- Change the SERVICE SELECTOR switch on the RT-662 or RT-834 to the FSK position.
- Change the RECEIVE switch on modem MD-522 from normal to reverse.
- Change the MODE SELECTOR switch on MD-522 to 850 Hz.
- Adjust the BFO on the MD-522 for reverse scope alignment when receiving a teletypewriter signal. If necessary, adjust the frequency vernier to assist BFO scope alignment when tuning the receiver to the receive signal.
- Conduct normal RATT operation.

Table 3-6. RATT tuning procedures, AN/GRC-46, VSC-1, VRC-29.

AN/GRC-46, VSC-1, VRC-29	AN/GRC-142/122, VSC-2, VSC-3
1. Change the SERVICE switch on the CV-278 from normal to reverse	1. Change the SERVICE switch on the RT-662 or RT-834 to FSK position.
2. Adjust the R-392 to the tuning signal of the AN/GRC-142/122, VSC-2 or VSC-3. Adjust until a mark 40 signal to the right of 0 is received on converter CV-278.	2. Change the RECEIVE switch on the MD-522 from normal to reverse
3. Tune the transmitter T-195 for a mark 40 to the right of 0 on the converter.	3. Change the MODE SELECTOR switch on the MD-522 to 850Hz
	4. Adjust the BFO control on the MD-522 for reverse scope alignment. If necessary, adjust the frequency varrier to assist the BFO scope alignment when tuning the receiver to the receive signal.

NOTE: When receiving a teletypewriter signal from like equipment (new series radio to new series radio), the RECEIVE switch must go back to NORMAL in order to receive.

(2) For old series RATT set AN/GRC-46, VSC-1, and VRC-29--

- Tune RATT equipment as usual for normal RATT mode of operation (Table 3-6).
- Change the SERVICE switch on converter CV-278 from normal to reverse.
- Adjust receiver R-392 to the tuning signal of the AN/GRC-142/122, VSC-2, or VSC-3. Adjust until a mark 40 signal to the right of 0 is received on converter CV-278.
- Realign transmitter to receiver.
- Conduct normal RATT operation.

(3) For old series RATT set AN/GRC-26D--

- Tune RATT equipment as usual for normal RATT mode of operation (Table 3-7).
- Change the MARK HOLD switch on converter CV-116 from XTAL (left) (NORM) position to the XTAL (right) (REV) position.
- Adjust receiver R-390 to the tuning signal of the AN/GRC-142/122, VSC-2, or VSC-3. Adjust until a mark 50 signal to the right of 0 is received.
- Realign transmitter to receiver.
- Conduct normal RATT operation.

Table 3-7. RATT tuning procedures, AN/GRC-26D.

AN/GRC-26D	AN/GRC-142/122, VSC-2, VSC-3
1. Change the MARK HOLD switch on the CV-116 converter from XTAL(left)(NORM) to XTAL(right)(REV).	1. Change the SERVICE SELECTOR switch on the RT-662 or RT-834 to the FSK position.
2. Adjust the R-390 to the tuning signal of the AN/GRC-142/122, VSC-2 or VSC-3. Adjust until until a mark 50 signal to the right of 0 is received on the CV-116 converter.	2. Change the RECEIVE switch on MD-552 from normal to reverse.
3. Tune the T-358 for a mark 50 reading on the right of 0 on the converter.	3. Change the MODE SELECTOR switch on the MD-522 to 850Hz.
	4. Adjust the BFO control on the MD-552 for reverse scope alignment. If necessary, adjust the frequency vernier to assist BFO scope alignment when tuning the receiver to the receive signal.
NOTE 1: When receiving teletypewriter signal from like equipment (old series radio to old series radio), the SERVICE switch on CV-278 or the MARK HOLD switch on CV-116 must go back to normal operating position to receive.	
NOTE 2: When operating secure RATT and you are not receiving, the teletypewriters may run open. To stop them from running open, return the SERVICE switch on the converter CV-278 to MARK HOLD position. On the CV-116, set MARK HOLD switch back to MARK HOLD position.	

Chapter 4

Telecommunications Center, Switching, and Patch Panel Operations

4-1. General

A signal organization must improvise when it does not have enough TCC equipment. All TCCs and RATT assemblages that currently provide over-the-counter record traffic support will be phased out by 1994. They will be replaced by user-owned and -operated LDF AN/UXC-7 and the microcomputer communications terminal AN/UGC-144. The user-owned and -operated operational concept/architecture will be phased in at EAC as the TRI-TAC Block III equipment (AN/TTC-39A and DGM) are fielded. AT ECB, phase in will be synchronized with the MSE fielding schedule.

4-2. Telecommunications Center Operations

When a signal organization is short of TCC equipment, it must use available shelters or properly secured and guarded tents. However accomplished, the TCC function must be performed.

a. Fabricate a TCC using spare shelters/tents, field tables, and spare teletypewriter equipment from inoperable RATT or TCC rigs. Be sure to use the appropriate on-line cryptographic equipment. This practice is subject to the command's policies on use and modification of equipment.

b. Establish a direct wire circuit between the TCC and the RATT rig handling priority traffic. Remoting the teletypewriter allows page copy to be transmitted and received directly at the TCC without additional processing or handling by radio operator personnel.

c. Cross train staff section clerical personnel in message preparation so that message traffic can be prepared in proper format for transmission before it is processed at the TCC. This lessens the impact of a shortage of TCC resources by spreading the workload. It does place an additional training burden upon the unit by requiring more people to know how to prepare messages. Messages must be short to optimize the use of available traffic channels. (See the discussion of low-level encryption and brevity codes in Chapter 7.) TCC personnel are not expected to modify or shorten messages by applying brevity codes to the messages. TCC personnel transmit message texts exactly as they receive them; any modification or shortening of messages must be accomplished by originators. (See DA Pam 25-7 for procedures on JINTACCS message text format.)

d. Increase the quantity of air and motor messengers available to make up for lack of TCC processing Switching, and Patch Panel Operations and transmission facilities. (See Chapter 6.)

e. Obtain and use additional quantities of AN/TCC-14 or AN/TCC-29 assemblies (which provide a speech-plus teletypewriter capability) for better use of available voice channels. Use of off-line encryption methods increases if TCC equipment is in short supply. (Low-level encryption methods are discussed in Chapter 7.)

NOTE: When assembling or fabricating temporary TCC facilities, it is essential to use proper grounding procedures.

f. Use facsimile equipment or microprocessors to route messages and overlays to their destinations. These devices can be used over different means of communications, such as FM radio, multichannel, and other facilities.

4-3. Telecommunications Center Operations Example

a. Your brigade CP area received enemy artillery fire, seriously injuring one operator and destroying the TCC equipment.

b. What is a temporary solution pending replacements for the TCC equipment?

c. One solution is to use a RATT rig as a temporary TCC. This requires several measures not normally used.

(1) The RATT rig chosen cannot operate in its normally assigned net full time. Provisions must be made for the passage of its usual traffic by alternate means or the passage of reduced amounts of traffic by the RATT rig for limited periods of time.

(2) The RATT equipment operators involved must have appropriate ACPs, other required publications, and the training to properly process the traffic.

(3) Additional personnel must be tasked and trained ahead of time to augment the message processing capability of RATT equipment operators.

(4) Maximum use of alternate means must be made during the shortage of the TCC equipment and personnel.

(5) Other equipment with teletypewriter capabilities can be tasked to perform TCC functions. This also requires advance cross training of additional personnel in TCC procedures.

(6) Originators must make special efforts to keep messages short during the period of the TCC shortage. Message brevity is a good practice any time, but is especially valuable during periods of equipment and personnel shortages.

4-4. Tactical Facsimile Operations

TCCs are rapidly being replaced by facsimile devices that can operate with the current inventory of communications equipment. These devices are user-operated and user-installed. The user can call up the addressee, confirm the link, transmit the copy, and verify the quality without interfacing with over-the-counter TCC service. This system has been fielded in Europe and virtually eliminates record traffic at brigade and separate battalion levels.

4-5. Switching Operations

Switching equipment is in short supply in many RC units and can also be expected to be in short supply during combat operations. This presents a serious problem to all commanders and communicators; however, measures can be taken to reduce the effect of switching equipment shortages.

a. Reduce the number of local lines to various subscribers. This must be done based on the commander's established priority system considering the minimum critical needs of the unit.

b. Establish a limited number of common-user telephone points or booths in staff areas. Phone usage in these activities should be restricted to certain predesignated users.

c. Limit lower priority calls to lesser-traffic hours and limit the call length. Noncritical administrative and logistics traffic can be passed during these hours or passed over alternate means.

d. Establish hot loops within specific activities and use a ringing code to alert users. This party line approach eliminates each party being connected directly to a switchboard and reduces the switchboard load.

e. Reduce manual operator interventions. Patch (or direct wire) priority circuits from the TOC and certain other priority users directly through the multichannel radio system (or alternate means) supporting that circuit to the distant subscriber.

f. Enforce telephone discipline during critical periods. Develop a local minimize policy for use over voice communications facilities.

g. Limit sole-user circuits to two-wire configurations and eliminate sole users in four-wire patterns.

h. Use local commercial telephone systems when possible.

4-6. Patch Panel Operations

The patch panel is literally the heart of a signal center operation. The absence or loss of a patch panel presents a large obstacle to the communicator but not an impossible one. Patching service can be provided when short a patch panel.

a. Fabricate a patching facility using distribution boxes J-1077A/U. Additional J-1077A/Us are essential in overcoming the problems of a missing patch panel. A cargo trailer can be used for this purpose. Mount J-1077A/Us on boards on each side of the trailer and fabricate an operator's table at the front. (See Figure 4-1.)

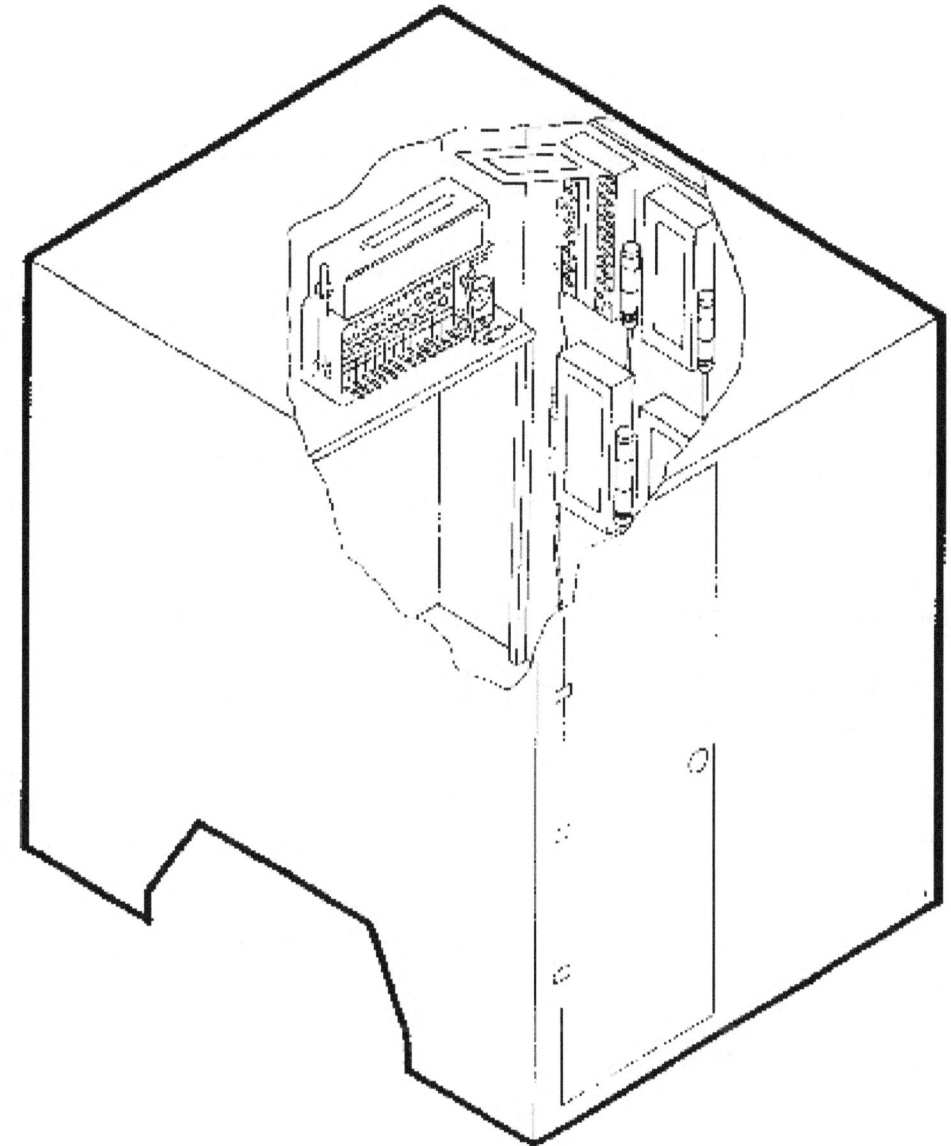

Figure 4-1. Cut-a-way view of fabricated patch facility.

(1) Maintain polarity when patching between J-1077A/Us using single strands of WF-16 field wire and labeling each connection.

(2) Develop a detailed patching log to control patching. This log would be substantially different from the normally used log due to the nature of the homemade patch facility.

(3) Mount a switchboard and phone on the operator's table for local interconnection and for circuit control and monitoring.

b. Use distribution box J-2317A/U or terminal box TA-125 as an alternative to using J-1077A/Us. The TA-125 cannot be directly connected to 26-pair cable but does provide a flexible method of interconnecting circuits. Short runs of WF-16 can be made from the terminals of the signal entrance boxes of various communications assemblages to the locally constructed patching facilities using distribution boxes J-1077A/U or J-2317A/U, or terminal box TA-125.

c. Use additional SB-22 switchboards, if available, to patch circuits on a limited basis. Their limited capacity severely restricts their use as patches; but, when used in conjunction with carefully controlled direct wiring between assemblages, they provide a measure of flexibility.

d. Have units that support a standard troop structure develop a prepatched board to handle known requirements. This leaves other limited patching facilities to handle new or changing requirements.

4-7. Patch Panel Replacement Example

a. You are a platoon leader. Enemy action has destroyed the patch panel. There are no casualties since the operators had time to take cover before the patch panel was hit. Replacements for the damaged cables are available.

b. How do you get your circuits operational pending receipt of another patch panel?

c. A possible solution is to establish a temporary patching facility using junction boxes such as J-1077A/U or J-2317A/U.

(1) Junction Box J-2317A/U is preferred since it terminates four each 26-pair cables and is not wired normal through as is the J-1077A/U. Consequently, the J-2317A/U is highly useful as a temporary patch facility.

(2) Patching is accomplished using lengths of field wire. Special patch records must be made to record these unusual field wire patches.

(3) Special attention must be given to patching the critical circuits first. Your systems control records will indicate these.

NOTE: Special attention must be given to maintaining proper polarity (in addition to RECEIVE-SEND pair transposition) when patching using this method.

Chapter 5

Wire and Cable Operations

5-1. General

a. Wire and cable are used to interconnect activities within CPs and between radio relay terminals and switching centers. Long haul wire circuits (trunks) are installed to complement radio systems when time, personnel, and equipment are available. Wire is especially useful in operations where movement is limited.

b. Summarized below are some of the advantages and disadvantages of using wire and cable.

    (1) Advantages.
        (a) Reduces the need for radio and decreases radio interference.
        (b) Reduces the electronic signature of CPs.
        (c) Reduces the enemy's jamming, interference, and direction-finding capabilities.
        (d) Provides backup and increased traffic-passing capabilities for radio systems.
        (e) Is not subject to interference or jamming.
        (f) Can be secured by VINSON or PARKHILL COMSEC equipment and wire line adapters.

    (2) Disadvantages.
        (a) Slow to install/recover. Requires additional manpower.
        (b) Not a secure means unless encrypted. Wire has some security but is subject to disruption.
        (c) Not reactive to fast-moving situations.
        (d) Limited by terrain and distance considerations.
        (e) Susceptible to damage by friendly action; for example, wheeled and tracked vehicle movement.
        (f) Susceptible to damage by indirect fire.
        (g) Is a good conductor of EMP which will damage attached telephone and switching equipment.

5-2. Installation Considerations

One of the first actions when establishing a signal site is the start of intra-site wire/cable installation. Practically all signal equipment at some point interfaces with wire/cable, especially when there is a shortage of multichannel equipment. The following actions are essential for successful wire/cable installation:

a. Plan for and requisition sufficient quantities of wire/cable and installation equipment to provide for known and recurring requirements. The range of field wire circuits varies with the type of wire and the terminating equipment. For planning purposes, the range of field wire circuits using battery operated telephones is 22.5 to 35.4 km (14 to 22 miles); using sound powered telephones, it is 6.4 km (4 miles).

b. Cross train personnel on wire/cable installation and operations. Personnel associated with equipment in short supply, and personnel in staff sections, should be cross trained so they will be capable of installing wire/cable circuits for their element/headquarters. Then each section that requires a telephone line can lay its own field wire to a centrally located, premarked distribution box J-1077A/U.

c. Plan for and install pre-positioned wire termination sites to the maximum extent possible. This pre-positioning should be related to planned or potential operations in order to provide headquarters elements immediate connection to the division/corps wire system upon displacement. This assumes that the potential site is in friendly territory and that operations in the area will not destroy the system. Typical uses of pre-positioned wire systems include quick connection of a brigade or tactical CP element or of a retransmission/NRI station to the system.

d. Fabricate short runs using WD-1 field wire when multiple-pair cables are in short supply. Use special caution to tag and identify pairs when using this technique.

e. Plan for additional vehicle, fuel, and maintenance support needed for IOM of wire and cable systems. Within most signal TOEs, the volume of cable and wire authorized exceeds the unit's capabilities to transport it in a single movement.

5-3. Equipment Utilization

a. The following wire and cable are used predominantly throughout the tactical arena:

(1) WD-1/TT consists of two 23-gauge conductors, individually insulated and twisted together.

(2) WD-lA/TT has two insulated conductors bonded together.

(3) WD-36/TT is a two-conductor, lightweight assault wire. It is used when rapid installation and light weight are factors of primary importance. It is designed for one-time use only.

(4) WF-16/TT is a four-conductor (two-pair) field wire that is used with the new 4-wire telephone system.

(5) WF-8( )/G, spiral-four cable, is a 4-wire transmission line for carrier communications systems. It is used for long distance VF circuits. A universal connector plug is attached to each end.

(6) CX-11230G is an interarea coaxial cable. It consists of two twisted coax tubes and provides transmission lines for 12-, 24-, 48-, and 96-channel TDM/PCM systems.

(7) CX-4566 is an intersite, multi-pair cable. It consists of 26-pair and is used for internal site connections and limited distance extensions to CPs.

b. Additional equipment is available to combat distortion and loss on long cable runs. TD-204 telephone repeaters are used to increase the strength of a signal that has been decreased by line loss. It consists of amplifiers and associated components (repeating coils, equalizer networks, and hybrid coils). TD-206B/G pulse form restorers retime and regenerate a pulse train on a PCM cable. The pulse form restorers are placed at one mile intervals up to a total of 39 restorers on a given cable run, for a total maximum of 64.4 km (40 miles). The maximum channel capability of the PCM cable run is 48 channels. The TD-202 in addition to TD-204 and TD-206B/G are used by AN/TRC-117, AN/TRC-110, AN/TCC-60/69, AN/TCC-61, and AN/TCC-65.

c. Fiber-optic cable has become an integral part of the Army inventory. It is available for use in the tactical environment. Fifteen-channel, fiber-optic cable will provide full duplex digital communications over a maximum range of 6 to 8 km (3.8 to 5 miles) with an average distance estimated to be 4 km (2.5 miles) (without restorers) at a data rate of 72 kilobits. With restorers, range, and rate increase up to 19.2 megabits and up to 64.4 km (40 miles).

5-4. Installation Methods

a. Aerial (overhead) wire/cable provides the most satisfactory service and is the easiest to maintain. However, it is subject to weather, enemy action, EMP, and artillery airbursts, and requires more time to install.

b. Surface wire/cable requires minimum time and equipment to install. Installed properly, it provides reliable circuits; installed improperly, it requires immediate and constant maintenance. It is also vulnerable to damage by troop, vehicular movement, and weather.

c. Buried wire/cable is more electrically stable than surface or aerial installation, and is less subject to weather. It is the preferred installation if time permits. It is difficult to maintain and recover. Buried cable also provides better protection from the effects of EMP in a nuclear environment.

5-5. Existing Wire/Cable Systems

Evaluation of available commercial cables may be done in peacetime for contingency use in war. Examples of systems that can offer alternative means are highway telephones, railway systems, and forestry services.

Chapter 6

Alternate Means of Communications

6-1. General

Numerous alternate means of communications are available to the field commander. The most commonly used will be addressed in the following paragraphs.

6-2. Messenger Service

Messenger service is one of the oldest and most effective alternate means of communications but is no longer a dedicated Signal Corps responsibility. Commanders can greatly enhance their ability to control their units by effectively using all types of messengers from all battalions and staffs.

a. The need for messengers is inversely related to our desire and/or ability to use radio. There are many disadvantages in relying too heavily on radio communications. Messenger service is an important consideration for combat operations. With the EMP effect becoming more of a tactical probability than a possibility, messenger service may become the only viable communications after the initial outbreak of wartime hostilities.

    (1) Advantages to messenger service are as follows:
        (a) Permits delivery of lengthy, bulky items not requiring immediate action; or requiring lengthy transmission times, such as, operation plans, map overlays, administrative/logistical matters, and reports.
        (b) Reduces the need for using radio.
        (c) Reduces mutual radio interference by reducing radio traffic and transmission time.
        (d) Reduces the enemy's EW capability by--
          • Providing a secure communications means which is not subject to EW.
          • Deferring exposure of critical nets until they become urgently needed.
          • Making it more difficult for the enemy to identify critical nets.

    (2) Disadvantages to messenger service are as follows:
        (a) Slower than electrical means.
        (b) Subject to enemy action, terrain, and weather conditions.
        (c) Requires dedicated equipment and trained personnel.

b. The type of messengers employed is determined by the urgency, bulk of messages, the terrain, the weather, and the availability of transportation.

    (1) Foot messengers are used for CP distribution and in small units.
    (2) Motor messengers are used between headquarters.
    (3) Air messengers are used to speed up delivery over long distances, over difficult terrain, over

enemy controlled territory, or for vital or urgent messages.

c. The types of service available by messengers, usually at EAC, are scheduled, special, and exchanged.

(1) Scheduled messengers follow prearranged and published time schedules, travel designated routes, and stop at predesignated points and headquarters. They normally deliver and pick up message traffic at TCCs or exchange points.

(2) Special messengers are used when scheduled service has not been established or to augment scheduled service. They are employed when the urgency of a message will not allow the message to wait for scheduled messenger service or transmission by other communication means.

NOTE: Couriers transporting COMSEC or classified materials orders IAW AR 66-5.

(3) Exchanged messengers are exchanged between specified headquarters when personnel and vehicle assets permit. These messengers would know routes back to their parent headquarters and would deliver high priority messages only.

d. Additional sources of messenger service should be considered.

(1) Personnel should never leave a CP en route to another unit without consulting the TCC for possible message traffic for that unit. Liaison officers can assist during times of limited messenger resources.

(2) Certain designated headquarters act as clearing houses and reroute points for messages. This allows unit messengers to pick up and deliver to these points, freeing dedicated messengers for more frequent delivery and pickup between relay points.

(3) Commanders should identify those organic assets which in an emergency could be used for messenger purposes. They should provide necessary training and planning so these assets can be effectively used when needed. All assets should be considered for the role; for example, the commander's vehicle and driver and the S3's vehicle and driver.

6-3. Visual Signaling

An alternative to electronic communications is visual signaling. This form of communications has been used extensively throughout the history of military operations. However, the flexibility, range, and speed of today's radios have reduced the use of visual signaling. Few units today would be capable of establishing a working system of visual communications without first conducting extensive planning and training.

a. Visual signaling systems normally require simple equipment. They provide timely point-to-point communications over the distances usually associated with company, battalion, and sometimes brigade and division deployments. They are not normally susceptible to EMC and EMP problems and are relatively reliable in many combat situations. Training requirements vary with the systems used and can be taught at unit level.

b. Visual signaling is not a cure-all. Numerous limitations need to be recognized. Visual signaling sites must be within LOS of each other and the signal means used must be distinguishable at the desired range. Fog, rain, snow, smoke, and background light conditions can reduce effective ranges.

c. Visual signaling cannot adequately handle the mass of routine communications between echelons, but it can be used as an alternate means to pass the high priority messages that may affect the tactical situation. Units should consider their requirements for continuous command communications, then develop a visual system that will provide an effective backup to their current systems. Realistic training of personnel and frequent practice during training exercises enhance a commander's ability to maintain effective control of assigned and attached units.

d. Visual signaling systems employing flashing lights of various types are most effective because they provide distinguishable signaling at great ranges. Several field expedient visual signaling devices are readily available to the tactical command.

> (1) Any standard flashlight can be pulsed in a controlled manner and can normally be seen over several hundred meters in daylight and up to two or more kilometers at night. A more directional beam can be obtained by attaching any sort of cylindrical extension.

> (2) Any other light source, including chemical lights, that can be keyed is also usable. An example is a vehicle headlight used with a keying device and a director to provide long-range signaling.

e. Reception of visual signals can be enhanced using readily available devices.

> (1) A pipe or other device pointed directly at the visual source can be used to limit the receiver's susceptibility to distraction. Such a device should be stabilized to prevent movement while in use.

> (2) Binoculars assist in distinguishing the signals at greater ranges. At night, the use of ambient light devices such as night vision goggles AN/PVS-5 or night vision sight AN/PVS-4 can greatly extend the system's range. The PVS-5 or PVS-4 is authorized in many divisional units including infantry/mechanized, infantry battalions, engineer battalions, division artillery units, cavalry and air cavalry troops, and air defense artillery battalions.

f. A flashing light system can be both an asset and a liability.

> (1) Flashing light signaling sites will often be visible from enemy positions and appropriate safeguards must be planned. Using infrared or near infrared sources and receiving devices will prevent unaided observation; however, most potential adversaries also use the infrared spectrum and can observe these signals.

> (2) Visual signaling sites should be remoted from CP locations to maintain CP location security. Remoting may also be required to attain LOS.

> (3) Portability and weight of equipment used is an asset. If the equipment can readily be mounted in trees or on antenna towers, LOS may be more easily attained.

(4) Communications between the signal site and CP is mandatory. Wire is strongly recommended, but short-range radio or a messenger could be used.

(5) Messages prepared for transmission by visual means should be as short as possible to facilitate the slower transmission speed and concentration required to copy at long ranges. Properly encrypted brevity codes should be used.

(6) Flashing light systems can use directional or nondirectional devices. Nondirectional equipment will generally provide less range, but can be used to transmit to more than one receiver simultaneously.

(7) All signal sites should be manned by a minimum of two personnel to enable one to focus attention on the distant sender and the other to record or relay the message over the telephone as it is received.

(8) Operators must be trained in the transmission and reception of code.

6-4. Panel Signaling

Panel signaling is used primarily between ground troops and aircraft. Its use is very limited and is covered by instructions in the unit's SOI. Care must be exercised in panel use to ensure that the panels can easily be seen from aircraft and that the hovering aircraft do not disclose CP or signal site locations when reading the panels.

6-5. Pyrotechnic Signaling

Pyrotechnic signaling uses colored smoke, colored and combinations of star clusters, or parachute flares. This kind of signaling is limited to prearranged signals. These signals are prescribed at command-, corps-, division-, brigade-, battalion-, and, under certain conditions, company-level. Meanings of signals are provided in the SOI or operation orders. Continual reassessment for completeness and adequacy of this signal system is encouraged.

Chapter 7

Communications Security Operations

7-1. General

a. Success on today's battlefield depends on the commander's ability to concentrate superior combat power at critical times and places. A key to this success is superiority in command and control via communications. Effective communications is essential to winning.

b. The enemy realizes the importance of our communications systems and will continuously try to interfere with our ability to communicate. He will try to gather intelligence from our communications, and then he will try to disrupt them. He will attempt to interfere by breaking into our nets and trying to deceive us, or he will try to jam us. Failing at these measures, he will try to destroy communications by fire. Our battlefield success will depend heavily on how well we minimize his attempts to disrupt our communications systems.

c. Communications security is the protection resulting from the application of crypto security, transmission security, and emission security. These protective measures are taken to deny unauthorized persons telecommunications information. This chapter addresses primarily the cryptographic and transmission security portions of COMSEC.

d. Presently, AC and RC have a substantial shortage of COMSEC equipment. No quick or easy solutions are in sight. The need for COMSEC is essential; if adequate quantities of COMSEC equipment are not available, the commander must take other measures to secure his communications. Additionally, the active Army must also be skilled in the use of manual encryption techniques. Because most RC units possess little secure voice equipment, the active Army must anticipate transmitting and receiving traffic with the reserves using manual encryption techniques. Remember, manual encryption is also the backup for loss or failure of machine crypto systems. Therefore, all forces must maintain proficiency in the manual encryption area, regardless of interaction with other forces. Below are alternative methods and systems which can be used in lieu of on-line crypto systems. They present some difficulty when large volumes of traffic must be processed; however, these methods are essential to assure success and survivability on future battlefields.

7-2. Authentication Systems

a. An authentication system is designed to protect a communications system against the acceptance of fraudulent transmissions. Everyone who communicates in a tactical or strategic environment requires some method of authenticating. Good authentication practices contribute to combat survival and effectiveness, because they aid in establishing the validity of a transmission, message, or originator. All commanders must implement their use during training and actual operations.

b. Combat experience in Vietnam proved that IED by the enemy contributed to substantial numbers of casualties and caused many missions to fall short of desired results. Proper authentication procedures can prevent an enemy from posing as a friendly station. The enemy is adept at IED and needs only a moderate degree of skill to seriously affect our communications when we do not authenticate. A balance has to be struck so that effective communications is maintained without harassment of friendly communications. Guidance on the use of authentication systems is found in the unit SOI, ACP 122(D), AR 380-40, and TB 380-41.

c. IED is the easiest EW technique to counter. Authentication is one of the best means available to stop enemy IED efforts. Operators are required to authenticate when they--

- Suspect a transmission is from an enemy station operating in the net (deception).
- Direct a station to go to radio silence or to break that silence. (Self-authentication can be used if authorized by the SOI.)
- Are challenged to authenticate.
- Talk about enemy contact, give an early warning report, or issue any follow-up report.
- Transmit directions which affect the tactical situation such as "Move to..." or "Turn off the radio." (Conversely, they challenge any directives like these with a request to authenticate.)
- Cancel a message.
- Open the net or resume transmitting after a long period of silence.
- Transmit to someone who is under radio listening silence.
- Transmit a classified message in the clear.
- Transmit messages in the blind; that is, neither desiring nor expecting a reply.

d. Challenge if you are not sure that authentication is required. If a station takes more than 5 seconds to authenticate, rechallenge. Why 5 seconds? Because an enemy operator may try to contact another station and have it respond to that same challenge, thereby obtaining the appropriate reply to your challenge.

e. The two most commonly used authentication procedures are challenge-reply and transmission authentication. The main difference between the two is that challenge-reply requires two-way communications, whereas transmission authentication does not. Even though transmission authentication requires only one-way communications, it is neither as simple nor as flexible as challenge-reply. The challenge-reply procedure most often used has a more flexible application.

    (1) Challenge-reply.

        C12 THIS IS A06 OVER
        A06 THIS IS C12 OVER
        C12 THIS IS A06 TURN EAST AT X-RAY OVER
        A06 THIS IS C12 AUTHENTICATE HOTEL VICTOR OVER
        C12 THIS IS A06 I AUTHENTICATE WHISKEY OVER
        A06 THIS IS C12 WILCO OUT

    (2) Transmission authentication.

    NOTE: Transmission authentication is used only when it is impossible or impractical to use challenge-reply authentication.

J8C THIS IS B6A DO NOT ANSWER
TURN EAST AT CROSSROAD X-RAY
AUTHENTICATION IS VICTOR PAPA
I SAY AGAIN
J8C THIS IS B6A DO NOT ANSWER
TURN EAST AT CROSSROAD X-RAY
AUTHENTICATION IS VICTOR PAPA
OUT

7-3. Transmission Security

Several categories of tactical information require transmission security protection. These are listed in FM 34-62, Appendix C. It is important that you learn these categories.

7-4. Codes

a. A code is a language substitution system that transforms plain language of irregular length, such as words or phrases, into groups of characters of fixed length. A code has an underlying plaintext of variable length, whereas a cipher has an underlying plaintext of fixed length (see paragraph 7-5). The codes that you will use are usually found in your unit SOI packet.

b. Two types of codes are normally used in tactical communications: security codes and brevity codes (only as authorized). A code used to hide meanings from another party is a security code. A code used to shorten transmissions is a brevity code. A brevity code only shortens transmission; it does NOT provide security. It is referred to as a brevity list. The international Q and Z signals found in ACP 131(D) and the police 10-code signals are examples of brevity lists. Brevity lists must be used in conjunction with an approved code to provide security.

c. Most codes can be placed into one of three categories: numerical, operations, and special purpose.

(1) Numerical codes are among the simplest and most useful types of codes and are used to encode numbers. They are almost always digraphs (two-letter configurations) and are designed to protect intelligence bearing QUANTITATIVE portions of tactical communications, especially voice communications. They provide a short term tactical advantage when it is impossible or impractical to secure information to any greater degree. They are intended for use through the lowest operational levels. They can be used to protect the when, where, and how many in communications that might otherwise be unencrypted. (For example, they may be mixed with plain language, operating signals, or a brevity list.) Numerical code examples are given below.

(a) Hours designation:
Plaintext--Meeting time is 1530.
Code value --Meeting time is I set XXBKWG.

(b) Frequency designation:
Plaintext --Change frequency to 14990 kHz at 1600.
Code value --Change frequency to I set XYFXMPE at HITV.

(2) OPCODE, in contrast with numerical codes, can be used to encrypt the what, who, why, how, and how many--the QUALITATIVE information in messages. They DO NOT, however, provide adequate protection for information if mixed with plain language. OPCODEs have a vocabulary of usually 1,000 to 3,000 entries and generally use trigraphic (three-letter) code groups. OPCODEs are usually multipurpose or general in that they may be used to encrypt different sorts of information. OPCODEs are intended for use down through the lowest operational levels. Operations codes examples are given below.

      (a) A simulated message of tactical operation assignment:
            Plaintext--Continue on assigned mission.
            Code value--AAL.
      (b) A simulated message of tactical operation report:
            Plaintext--Battalion action at crossroad.
            Code value --OXW RFM RFX WOX.

(3) Special purpose codes are OPCODE-type items but are generally designed for encrypting specialized types of messages such as radar reports and fire missions. Their vocabularies are usually limited in size and scope and may consist of single letter, digraphic, or trigraphic code groups. Frequently, they are intended for more sensitive application than general codes and are often used at higher echelons. Many such special purpose codes are of the one-time variety. Special purpose code examples are given below.

      (a) Simulated message --Communications equipment, radar station reports:
            Plaintext--TRC capability impaired, jamming suspected hostile.
            Code value--OTV JNP.
      (b) Simulated message --Artillery mission report:
            Plaintext--Target altitude 2,500 meters.
            Code value--XHY OHT.

d. Many codes are custom designed to meet requirements of specific users. They are fabricated in response to specific COMSEC needs. They can be produced for any commander who requires an individually tailored item. These codes are not to be produced without NSA approval. Most users, however, use only a few standardized systems. Standardized systems can be obtained and pre-positioned at appropriate levels in the distribution. This should be a consideration when requesting COMSEC support. If not regularly used, codes may not be on hand at the local COMSEC support agency. If not readily available, codes must be ordered through COMSEC logistic channels. Users and commanders should consider that custom designed codes require sufficient lead time to produce. Although a standardized system may not be the best solution to a tactical COMSEC problem, it may serve as an effective interim system until a more suitable custom designed product can be produced. You should never try to make up your own brevity codes since experience has shown they are too easily broken by the enemy. Only use authorized and approved codes.

e. The use of codes to gain advantage over an enemy cannot be overemphasized. Everyone using codes must be familiar with their capabilities, limitations, and intended usages for codes to be effective.

(1) Codes intended for tactical application are designed to provide ONLY that amount of security consistent with operational needs.

(2) Tactical OPCODEs usually require that messages be composed prior to being encrypted and

transmitted. Users need a pencil, paper, and a place to write in order to work on OPCODEs.

(3) A tactical OPCODE is of specific but limited usefulness in the operational environment. It is difficult to use in the midst of hostilities or when riding in a vehicle. It cannot adequately protect high level communications.

(4) Numerical codes can usually be operated without pencil and paper. Numerical codes can provide protection to quantitative elements of information that pertain to an immediate tactical situation. However, they are not as secure as properly used OPCODEs or numerical ciphers.

(5) All codes have a cryptoperiod. They also have usage rules that outline restrictions on their employment. If a code is used in a way for which it is unintended, security can break down quickly. Total encryption using tactical codes is not always desirable or possible. Encryption of information the enemy already knows may help assist him in breaking our code system.

(6) One-time codes have special usage characteristics because of their one-time cryptoperiod. These codes provide a high degree of security and can be used for traffic with long-term intelligence value.

(7) The commander can request that the local INSCOM counter SIGINT personnel produce a code that will meet his needs when an emergency arises that does not allow a unit to use authorized codes (such as, compromise or cut off of distribution). Under no circumstances should unauthorized codes be used.

(8) Training must emphasize security and resupply procedures for codes to ensure that all personnel involved in their handling and use are properly trained.

7-5. Ciphers

a. The one-time pad is a language substitution cipher system which transforms plain language formations of fixed length (numbers and/or letters) into characters or groups of characters of fixed length. In a cipher system, the underlying plaintext is fixed in length; in a code system the underlying plaintext is variable in length. A one-time pad has no vocabulary as such, and almost anything can be said using pads. One-time pads are highly secure and are used mainly for special operations.

b. All one-time pads have variables which are used to transform plaintext into cipher text. These variables are presented in the form of recognizable characters such as letters and/or numbers. Each individual key is used only once, from which is derived the name one-time pad.

c. The substitution of cipher text for plaintext and vice versa is performed according to a specified rule which uses the key variables discussed above. The rule, how to work the pad, is what distinguishes one type of system from another.

d. The three basic varieties of one-time pads are literal, digital, and literal/digital.

(1) A literal pad can encrypt letters only, so numbers must be spelled out before encryption. This gives great flexibility in the variety of plaintext that can be encrypted, but also results in a

longer encryption time than would be experienced with a code.

(2) A digital pad encrypts digits only. If information to be protected is strictly numerical, digital pads can directly encrypt the plaintext. It is not uncommon for it to be used to directly encrypt narrative text if the text can be taken from a brevity list whose equivalent groups are numerical. This technique is especially valuable between speakers of different languages, as operators need no linguistic skills since transmission involves only digits.

(3) Literal/digital pads are used to encrypt both letters and numbers directly. Their applications are similar to literal-only pads except they can directly encrypt numbers without spelling them out. They are most useful over good quality circuits which are least likely to require spelling numbers.

(a) Standardization systems constitute the majority of one-time pads. Standard systems accommodate most operational systems, but custom designed pads can be ordered through the local COMSEC support agency by any commander who has a legitimate need for special material.

(b) Pads can be used to protect highly sensitive traffic since they provide security for an indefinite time. Pads require a pencil for their operation, and some require that messages be composed before encryption. Writing space is normally provided for writing directly on the pad. A pad key is intended for one-time use only. If more than one message is enciphered in the same stretch of key, it is possible to break both messages.

(c) One-time pads, like one-time codes, are most effective on point-to-point or broadcast nets.

(d) One-time pad example:
Plaintext--l4.5. Code value--XKBQ.
Message transmitted --Page 030, set 3, XKBQ.

e. Numerical ciphers, as with a pad, are characterized by the fixed length of the underlying plaintext. In all cases, this is a one-for-one substitution.

(1) The two types of numerical ciphers are the one-time ciphers and the standard cipher system. The one-time ciphers are an easy-to-employ, highly secure, numerical enciphering system. They can be used on basic numerical data or on a fixed format, such as specific data reports for personnel summaries. The standard numerical cipher system for enciphering numbers is DRYAD.

(2) A limited transmission authentication capability and a challenge/reply authentication capability are also provided with this system.

7-6. Brevity Lists

a. There may be instances where the types of information to be exchanged are not sufficiently varied to warrant the use of an extensive operations code. In such cases, it may be preferable to use a brevity list

(Figure 7-1) in conjunction with a numerical cipher. Security is provided by encrypting numerical equivalents with an approved cipher system such as the DRYAD system.

```
Execution
062 Execute aerial recon of (loc) to (loc) at (time).
063 Execute airborne assault at (loc), (time).
064 Execute airmobile assault
065 Execute ambush at (loc), (time).
…
089 Execute screen forward
090 Spare
091 Spare
092 Spare
```

Figure 7-1. Example of Brevity List.

b. Use of a brevity list has certain advantages over use of an operations code. Use of a brevity list eliminates the need to distribute, account for, and destroy an operations code in which the greatest part of the vocabulary is never used.

c. Brevity lists may be found in the supplemental instructions in the unit SOI. If there is no list present, it may be added at the unit level permanently or temporarily for a particular exercise.

d. A brevity list approach is an extremely practical alternative to an operations code--

- When the information exchange requirement is relatively limited.
- When the entries can be held to a minimum number of sentences, phrases, and/or words.
- When the messages are generally short.
- Where the vocabulary entries consist primarily of complete, independent thoughts.

7-7. COMSEC Operations Support

a. A unit must have a COMSEC account or have access to a COMSEC account, before it can conduct meaningful COMSEC training. Most units either have approved containers or a facility for storing COMSEC material or can obtain an approved container through supply channels. Command is responsible for establishing a COMSEC account. The unit must then train and operate using COMSEC equipment and/or systems. Commanders must use AR 380-40 and the TB 380-41 series in establishing COMSEC support for their unit. Although it places an additional burden on the commander to use COMSEC systems, their use is essential to success and survival on the battlefield. COMSEC IS NOT AN OPTION; IT IS MISSION-ESSENTIAL.

b. Unit SOPs must be clear on the use of COMSEC equipment and systems, both for administrative operations and tactical operations. All personnel should be familiar with SOP instructions and SOI instructions on the unit's particular COMSEC systems. Personnel must be familiar with procedures for resupply of COMSEC material during operations, and the COMSEC custodian must provide for an adequate supply of COMSEC material to be on hand for both training and/or operations.

c. Training programs must ensure that all necessary personnel receive adequate instructions and training on COMSEC procedures by both formal and on-the-job training. The INSCOM support activity can provide invaluable assistance in establishing, maintaining, and evaluating your unit's COMSEC account, training program, SOP, and storage facilities. Their support must be scheduled well ahead of time due to the number of units each activity supports. Information on storage and accounting for COMSEC equipment can be found in AR 380-40, AR 640-15, and the TB 380-41 series.

Chapter 8

Tactical Satellite Communications Operations

8-1. General

a. TACSATCOM helps to fulfill the need for command and control communications on the modern battlefield. TACSATCOM has the following assets and features:

- Extended range.
- High transmission reliability.
- Rapid emplacement.
- Easy siting.
- System flexibility.
- Survivability.
- Burst transmission and low power output capability.
-

b. These critical requirements are not fully satisfied by terrestrial communications systems. However, the Army is fielding GMF satellite communications that will complement communications links now served by LOS radio relay troposcatter and HF SSB radio systems. This satellite communications capability will greatly improve command and control communications.

8-2. Advantages

TACSATCOM systems are uniquely capable of meeting the above parameters. Their proficiency lends a broader scope to communications.

a. Range. The number of satellite terminals that can be supported in a theater of operation depends upon the location of each terminal with respect to the satellite antenna footprint, mode of operation, number of channels used, and condition/gain setting of the satellite itself.

b. Reliability. The TACSAT links will be equipped with organic antijam circuitry to enable them to survive certain degrees of intentional and unintentional radio interference. However, severe weather can cause satellite links to degrade.

c. Rapid emplacement. Set up and tear down of the ground terminal takes about 30 minutes.

d. Easy siting. High ground is not required; rather, natural terrain features, such as valleys, can be used to shield the terminal from detection or interference from ground-based emitters. However, masking of the antenna must be avoided. This will require level ground and may require low horizon and open area.

e. Flexibility. Since all TACSATCOM terminals in a given theater use the same satellite, connectivity can be quickly reconfigured providing great flexibility in changing battlefield conditions. The TACSAT link can provide continuous communications between widely dispersed elements of a highly mobile tactical force.

f. Survivability. The mobility of these terminals, large bandwidths, antijam capability, and siting advantages greatly reduce the threat to satellite terminals. However, if the satellite becomes inoperative for any reason, the TACSATCOM system is shut down.

8-3. Deployment

a. The terminals will augment selected HF, LOS, and tropo multichannel systems from brigades to theater Army. They will satisfy critical command and control multichannel transmission requirements from the maneuver brigade level through echelons above corps. In most cases, multichannel GMF/TACSATCOM will reduce the number of terrestrial LOS and tropospheric scatter terminals required to support a force, but will not eliminate the need for terrestrial multichannel communications. The terrestrial multichannel terminals will be used to support less critical and/or shorter communications links.

b. The AN/PSC-3 and AN/VSC-7 will be used by special forces and ranger units for minimum essential communications. The AN/URC-101 and AN/URC-110 are used for special contingency units at selected corps and division level. The AN/MSC-64 will be used primarily for emergency action record traffic in units required to be in the special communications system network.

c. The multichannel SHF system (AN/TSC-85A and AN/TSC-93A) will augment the multichannel systems at corps, division, and brigade level. For example, a TACSATCOM system will be assigned to a separate brigade, giving the separate brigade a fast moving, flexible, and reliable communications capability with its higher headquarters. For doctrinal deployment at division level, see FM 11-50.

Appendix

MSE Interoperability

A-1. Planning for MSE Interface

a. The requirement for establishing and controlling communications remains from higher to lower, left to right, and supporting to supported. With MSE that doctrine transcends more than just establishing and maintaining network integrity. The element in the higher, left, or supporting category also supplies the requisite equipment when augmentation is needed and coordinates and provides the necessary frequencies, frequency plans, COMSEC keys, codes, software, and control mechanisms.

b. MSE interface with other systems, such as EAC, TRI-TAC, or NATO, requires detailed planning and coordination. Signal planners must coordinate signal timing relationships, digital trunk group numbering and channel assignments, area codes, digit editing, and exchange of TG and AIRK COMSEC keys to ensure successful switch interface. Normally, the MSE gateway switch will modify its data base to accommodate the TRI-TAC switch.

A-2. Installation Parameters for MSE Interface

The CX-11230 cable used to interconnect the various assemblages is issued in 1/4-mile reels. Tables A-1 through A-7 give cable adjustment settings for lengths of cable from 1/4 to 1 mile (1 to 4 reels). The "cable reels" line lists the four possible cable lengths as 1/2/3/4 for 1, 2, 3, and 4 reels respectively. The cable adjustment for each cable length is similarly listed for transmit and receive at each assemblage. For example, a transmit listing of 4/4/4/4 means that the setting is 4 for each of the four possible cable lengths. The letters a, b, c, d, and e are used for settings of 0, 1/4, 1/2, 3/4, and 1 mile respectively. NA means no adjustment or not applicable.

Table A-1. NS to EAC via AN/TRC-151.

	NS	TRC-151	TRC-161	TTC-39/39A
Timing	Master	NA	NA	Master
Bit rate	576 kb/s	576 kb/s	576 kb/s	576 kb/s
Modulation	Dipulse	Dipulse	Dipulse	Dipulse
Cable reels	1/2/3/4	1/2/3/4	1/2/3/4	1/2/3/4
Cable xmit	4/4/4/4	e/e/e/e	e/e/e/e	4/4/4/4
Cable rcv	1/2/3/4	b/c/d/e	b/c/d/e	1/2/3/4
CCS channel	1st	NA	NA	1st
RSS channel	NA	NA	NA	NA
Traffic channels	2-32	NA	NA	2-32
Control node	No	No	Yes	Yes
Glare	Accept	NA	NA	Reject

Table A-2. NS to EAC via AN/TRC-170.

	NS	TRC-170	TRC-170	TTC-39/39A
Timing	Master	NA	NA	Master
Bit rate	512 kb/s	512 kb/s	512 kb/s	512 kb/s
Modulation	Diphase	Diphase	Diphase	Diphase
Cable reels	1/2/3/4	NA	NA	1/2/3/4
Cable xmit	4/4/4/4	NA	NA	4/4/4/4
Cable rcv	1/2/3/4	NA	NA	1/2/3/4
CCS channel	1st	NA	NA	1ST
RSS channel	NA	NA	NA	NA
Traffic channels	2-32	Na	NA	2-32
Control node	No	No	Yes	Yes
Glare	Accept	NA	NA	Reject

Table A-3. NS to EAC via AN/TSC-85A/93A (using TD-1337 TRI-TAC port).

	NS	TSC-93A	TSC-85A	TTC-39/39A
Timing	Master	CNCE	CNCE	Master
Bit rate	576 kb/s	576 kb/s	576 kb/s	576 kb/s
Modulation	Diphase	Diphase	Diphase	Diphase
Cable reels	1/2/3/4	1/2/3/4	1/2/3/4	1/2/3/4
Cable xmit	4/4/4/4	1/2/3/4	1/2/3/4	4/4/4/4
Cable rcv	4/4/4/4	1/2/3/4	1/2/3/4	4/4/4/4
CCS channel	1ST	NA	NA	1ST
RSS channel	NA	NA	NA	NA
Traffic channels	2-32	2-32	2-32	2-32
Control node	No	No	Yes	Yes
Glare	Accept	NA	NA	Reject

Table A-4. NS to EAC via AN/TSC-85A (using MD-1026).

	NS	TSC-85A	TSC-85A	TTC-39/39A
Timing	Master	CNCE	CNCE	Master
Bit rate	576 kb/s	576 kb/s	576 kb/s	576 kb/s
Modulation	Diphase	Diphase	Diphase	Diphase
Cable reels	1/2/3/4	1/2/3/4	1/2/3/4	1/2/3/4
Cable xmit	4/4/4/4	NA	NA	4/4/4/4
Cable rcv	4/4/4/4	NA	NA	1/2/3/4
CCS channel	1ST	NA	NA	1ST
RSS channel	NA	NA	NA	NA
Traffic channels	2-32	2-32	2-32	2-32
Control node	No	No	Yes	Yes
Glare	Accept	NA	NA	Reject

Table A-5. NS to NS via AN/TSC-85A/93A (using TD-1337 TRI-TAC port).

	NS	TSC-85A	TRC-93A	NS
Timing	Master	CNCE	CNCE	Master
Bit rate	1152 kb/s	1152 kb/s	1152 kb/s	1152 kb/s
Modulation	Diphase	Diphase	Diphase	Diphase
Cable reels	1/2/3/4	1/2/3/4	1/2/3/4	1/2/3/4
Cable xmit	4/4/4/4	1/2/3/4	1/2/3/4	1/2/3/4
Cable rcv	4/4/4/4	1/2/3/4	1/2/3/4	4/4/4/4
CCS channel	1ST	NA	NA	1ST
RSS channel	2d	NA	NA	2d
Traffic channels	3-64	NA	NA	3-64
Control node	Yes	Yes	No	No
Glare	Reject	NA	NA	Accept

Table A-6. NS to NS via AN/TSC-85A (using MD-1056).

	NS	TSC-85A	TRC-85A	NS
Timing	Master	CNCE	CNCE	Master
Bit rate	1152 kb/s	1152 kb/s	1152 kb/s	1152 kb/s
Modulation	Diphase	Diphase	Diphase	Diphase
Cable reels	1/2/3/4	1/2/3/4	1/2/3/4	1/2/3/4
Cable xmit	4/4/4/4	NA	NA	4/4/4/4
Cable rcv	1/2/3/4	NA	NA	1/2/3/4
CCS channel	1ST	NA	NA	1ST
RSS channel	2d	NA	NA	2d
Traffic channels	3-64	NA	NA	3-64
Control node	Yes	Yes	No	No
Glare	Reject	NA	NA	Accept

Table A-7. NS to LENS via AN/TSC-85A/93A (using TD-1337 TRI-TAC port).

	NS	TSC-85A	TSC-93A	LEN
Timing	Master	CNCE	CNCE	Master
Bit rate	576 kb/s	576 kb/s	576 kb/s	576 kb/s
Modulation	Diphase	Diphase	Diphase	Diphase
Cable reels	1/2/3/4	1/2/3/4	1/2/3/4	1/2/3/4
Cable xmit	4/4/4/4	1/2/3/4	1/2/3/4	4/4/4/4
Cable rcv	4/4/4/4	1/2/3/4	1/2/3/4	4/4/4/4
CCS channel	1ST	NA	NA	1ST
RSS channel	2d	NA	NA	2d
Traffic channels	3-32	NA	NA	3-32
Control node	Yes	Yes	No	No
Glare	Reject	NA	NA	Accept

Glossary

Abbreviations and Acronyms

AC- Active Components
ACP- Allied Communication Publication
AIRK- area inter switch rekey
AM- amplitude modulated/amplitude modulation
BFO- beat frequency oscillator
CEOI- Communications-Electronics Operation Instructions
CG- commanding general
CNCE- communications nodal control element
COMSEC- communications security
CP- command post
CW- continuous wave
DA- Department of the Army
DGM- digital group multiplexer
DISCOM- division support command
DSB- double side band
DTAC- division tactical command post
DTMF- dual-tone multifrequency
DTOC- division tactical operations center
DZ- drop zone
EAC- echelon above corps
ECB- echelons corps and below
ECCM- electronic counter-countermeasures
EMC- electromagnetic compatibility
EMP- electromagnetic pulse
EW- electronic warfare
FASC- forward area signal center
FDX- full-duplex
FLOT- forward line of own troops
FM- frequency modulated/frequency modulation
FSE- fire support element
FSK- frequency shift keying
G1- Assistant Chief of Staff, G1 (Personnel)
G2- Assistant Chief of Staff, G2 (Intelligence)
G3- Assistant Chief of Staff, G3 (Operations and Plans)
G4- Assistant Chief of Staff, G4 (Logistics)
GMF- ground mobile forces
GW- ground wire
HF- high frequency
HQ- headquarters
Hz- Hertz
IAW- in accordance with
IED- imitative electronics deception

IHFR- improved high frequency radio
INSCOM- United States Army Intelligence and Security Command
IOM- installation, operation and maintenance
JINTACCS- Joint Interoperability of Tactical Command and Control Systems
JUH-MTF- joint user handbook for message text format
kb/s- kilobits per second
KC- kilocycles (kilohertz)
kHz- kilohertz
km- kilometer
LDF- lightweight digital facsimile
loc- location
LOS- line of sight
MC- megacycles (megahertz)
MHz- megahertz
mi- miles
MODEM- modulation/demodulation equipment
MOPP- mission-oriented protection posture
MSE- mobile subscriber equipment
NA- not applicable
NATO- North Atlantic Treaty Organization
NBC- nuclear, biological, chemical
NCO- noncommissioned officer
NCS- net control station
norm- normal
NRI- net radio interface (see RWI)
NS- node switch
NSA- National Security Agency
NSK- narrowband shift keying
OPCODE- operations code
OPLAN- operations plan
OPORD- operation order
OPSEC- operations security
PCM- pulse code modulation
PEP- peak envelope power
RATT- radio teletypewriter
RC- Reserve Components
rcv- receiver
rec- receive
recon- reconnaissance
retrans- retransmission
rev- reverse
RF- radio frequency
RT- receiver-transmitter
RWI- radio wire integration (changed to NRI-net radio interface)
S1- Adjutant (US Army)
S2- Intelligence Officer (US Army)
S3- Operations and Training Officer (US Army)
S4- Supply Officer (US Army)
SCCS- satellite communications control system

SHF- super high frequency
SIGCEN- signal center
SIGINT- signals intelligence
SINCGARS- Single-Channel Ground/Airborne Radio System
SOI- signal operation instructions
SOP- standing operating procedure
SSB- single side band
STAJ- short term anti-jam
suppl- supplement
tac- tactical
TACFIRE- tactical fire direction system
TACSAT- tactical satellite
TACSATCOM- tactical satellite communications
TCC- telecommunications center
TOC- tactical operations center
TOE- table(s) of organization and equipment
TRADOC- United States Army Training and Doctrine Command
TRI-TAC- Joint Tactical Communications
VF- voice frequency
VHF- very high frequency
VRC- vehicular radio configuration
w- watt
xmit- transmit
XTAL- crystal

Terms

BREVITY LIST.
    A list used to shorten transmissions. It does not provide security.
DETENT TUNING.
    A method of tuning a radio to an operating frequency where the dial clicks into place at each frequency.
FIBER-OPTIC CABLE.
    A small, lightweight cable using pulses of light for transmission. Very low error rates but extremely susceptible to damage.
JAMMING.
    The intentional transmission of radio signals in order to interfere with the reception of signals from another station.
KEY LIST.
    A publication containing the key for a particular cryptosystem in a given cryptosystem.
MODULATE.
    To vary the amplitude, frequency or phase of a wave by impressing one wave on another wave of constant properties.
NESTOR.
    A communications security device.
NET.
    Organization of stations capable of direct communications on a common channel, often on a

definite schedule.

NET AUTHENTICATION.

Identification used on a communications network to establish the authenticity of several stations.

PARKHILL.

COMSEC device for HF radios.

PATCH PANEL.

A panel that contains means for changing circuit configurations; usually, it consists of receptacles/jacks into which jumpers/plugs can be inserted.

PATCHING.

Connecting two lines or circuits together temporarily by means of a patch cord.

POLARITY.

A condition by which the direction of the flow of current can be determined in an electrical circuit. Having two opposite conditions; one positive and one negative.

SECURITY CODE.

A code used to hide meanings from another party.

SIDEBANDS.

The frequency bands on both sides of the carrier frequency.

SQUELCH.

To shut off the audio output of a radio when a signal is not being received.

TRAINS.

A service force or group of service elements which provide logistics support.

TROPOSPHERIC SCATTER (TROPOSCATTER).

The propagation of radio waves by scattering, as a result of irregularities or discontinuities in the physical properties of the troposphere.

VINSON.

COMSEC device for combat net radios.

References

Required Publications

Required publications are sources which users must read in order to understand or to comply with this publication.

Army Regulations (AR)

AR 530-2 Communications Security
AR 530-3 (C) Electronic Security (U)

Field Manuals (FM)

FM 11-50 (HTF) Combat Communications Within the Division (How to Fight)
FM 24-1 Combat Communications

Related Publications

Related publications are sources of additional information. Users do not have to read them to understand this publication.

Allied Communication Publications (ACP)

ACP 122 (D) (C) Communications Instructions--Security (U)
ACP 125 (D) Communications Instructions--Radiotelephone Procedures
ACP 131 (D) Communications Instructions--Operating Signals

Army Regulations (AR)

AR 66-5 Armed Forces Courier Service
AR 105-64 US Army Communications- Electronics Operation Instructions Program
AR 310-25 Dictionary of United States Army Terms (Short Title: AD)
AR 310-50 Authorized Abbreviations and Brevity Codes
AR 380-5 Department of the Army Information Security Program
AR 380-40 (C) Policy for Safeguarding and Controlling COMSEC Information (U)
AR 530-1 Operations Security (OPSEC)
AR 530-4 (C) Control of Compromising Emanations (U)
AR 640-15 Criteria for Insuring the Competency of Personnel to Install, Maintain and Repair Communications Security Equipment

Department of the Army Form (DA Form)

DA Form 2028 Recommended Changes to Publications and Blank Forms

Department of the Army Pamphlets (DA PAM)

DA PAM 25-7 Joint User Handbook for Message Text Formats (JUH-MTF)
DA PAM 25-30 Consolidated Index of Army Publications and Blank Forms

Field Manuals (FM)

FM 3-3 NBC Contamination Avoidance
FM 3-4 NBC Protection
FM 3-5 NBC Decontamination
FM 3-100 NBC Operations
FM 24-2 Radio Frequency Management
FM 24-16 Communications- Electronics: Operations, Orders, Records and Reports
FM 24-17 Tactical Telecommunications Center Operations
FM 24-18 Tactical Single-Channel Radio Communications Techniques
FM 24-33 Communications Techniques: Electronic Counter-Countermeasures
FM 24-35 (0) Communications-Electronics Operations Instructions (The CEOI)
FM 24-35-1 Signal Supplemental Instructions
FM 34-62 Counter- Signals Intelligence (C-SIGNT) Operations

Technical Bulletins (TB)

TB 380-41 (0) Procedures for Safeguarding, Accounting, and Supply Control of COMSEC Material

Technical Manuals (TM)

TM 11-5135-15 Radio Set Control AN/GSA-7
TM 11-5820-398-12 Operator's and Organizational Maintenance Manual (Including Repair Parts and Special Tool Lists): Radio Set, AN/PRC-25 (Including Receiver Transmitter, Radio, RT-505/ PRC-25)
TM 11-5820-401-10-1 Operator's Manual for Radio Sets AN/VRC-12, AN/VRC-43, AN/VRC-44, AN/VRC-45, AN/VRC-46, AN/VRC-47, AN/VRC-48, and AN/VRC-49 (Used Without Intercom Systems)
TM 11-5820-401-10-2 Operator's Manual: Radio Sets, AN/VRC-12, AN/VRC-43, AN/VRC-44, AN/VRC-45, AN/VRC-46, AN/VRC-47, AN/VRC-48 and AN/VRC-49 (Used With an Intercom System)
TM 11-5820-401-20-1 Organizational Maintenance for Radio Sets, AN/VRC-12, AN/VRC-43, AN/VRC-44, AN/VRC-45, AN/VRC-46, AN/VRC-47, AN/VRC-48 and AN/VRC-49 (Used Without Intercom Set)
TM 11-5820-401-20-2 Organizational Maintenance for Radio Sets, AN/VRC-12, AN/VRC-43, AN/VRC-44, AN/VRC-45, AN/VRC-46, AN/VRC-47, AN/VRC-48 and AN/VRC-49 (Used With Intercom System, AN/VIC-1(V))

TM 11-5820-498-12 Operator's and Organizational Maintenance Manual: Radio Sets, AN/VRC-53, AN/VRC-64, AN/GRC-125 and AN/GRC-160 and Amplifier-Power Supply Groups OA-3633/GRC and OA- 3633A/GRC

TM 11-5820-667-12 Operator's and Organizational Maintenance Manual: Radio Set, AN/PRC-77 (Including Receiver Transmitter, Radio RT-841/PRC-77)

Training Circulars (TC)

TC 24-20 Tactical Wire and Cable Techniques
TC 24-21 Tactical Multichannel Radio Communications Techniques
TC 24-24 Signal Data References: Communications-Electronics Equipment

Projected Publications

Projected publications are sources of additional information that are scheduled for printing but are not yet available. Upon print, they will be distributed automatically via pinpoint distribution. They may not be obtained from the USA AG Publications Center until indexed in DA Pamphlet 25-30.

Field Manuals (FM)

FM 11-50 Combat Communications Within the Division (Heavy and Light)
FM 24-35 Signal Operation Instructions: The "SOI"